T0305724

Think RISK

Risk is the single most prevalent and enduring factor that influences every individual, organization, and society. People often seek protection from negative risk events, but also seek to take advantage of opportunities arising from positive risk events. We may feel overwhelmed by messages encountered in daily interactions with media and society, contributing to a sense of ambiguity over how to act in response to risk-related information and misinformation. We seek to leverage evidence and reason to find our own balance between both positive and negative outcomes in an uncertain world. This groundbreaking book delivers practical concepts and tools that empower readers to leverage innovations in risk science to improve their abilities to interpret, assess, communicate, and handle risk. It provides a practical non-quantitative approach to understanding the risk and making better decisions involving risk.

Think RISK covers several key themes in risk science: (a) the main goals and strategies for understanding and managing risk; (b) how readers can inform their risk stances by considering their own individual values and mission; (c) the difference between risk and safety, and how that difference is critical for managing the risk; (d) the role of psychological factors when understanding and managing the risk; (e) the role of communication when understanding and managing the risk; and (f) the general importance and incentives for effectively understanding and managing the risk.

Written for business professionals in all private and public sectors, this book will also be relevant to non-business professionals such as medical practitioners and policymakers and would be an ideal fit for executive education and seminar-style courses in universities, corporate book clubs, and training seminars. Because it's based on foundational and scientifically accepted ideas and principles, the book should remain relevant for many years.

Shital Thekdi is Associate Professor of Analytics and Operations at the University of Richmond. She has coauthored several papers on risk management and decision-making.

Terje Aven is Professor of Risk Analysis and Risk Management at the University of Stavanger, Norway, since 1992. Previously he was also Professor (adjunct) of Risk Analysis at the University of Oslo and the Norwegian University of Science and Technology.

Think RISK

A Practical Guide to Actively Managing Risk

Shital Thekdi and Terje Aven

LONDON AND NEW YORK

First published 2023
by Routledge
4 Park Square, Milton Park, Abingdon, Oxon OX14 4RN

and by Routledge
605 Third Avenue, New York, NY 10158

Routledge is an imprint of the Taylor & Francis Group, an informa business

British Library Cataloguing-in-Publication Data
A catalogue record for this book is available from the British Library

Library of Congress Cataloging-in-Publication Data
Names: Thekdi, Shital, author. | Aven, Terje, author.
Title: Think risk : a practical guide to actively managing risk / Shital Thekdi and Terje Aven.
Description: Abingdon, Oxon ; New York, NY : Routledge, 2023. | Includes bibliographical references and index.
Identifiers: LCCN 2022056526 | ISBN 9781032358925 (hardback) | ISBN 9781032358901 (paperback) | ISBN 9781003329220 (ebook)
Subjects: LCSH: Risk management. | Decision making. | Risk assessment.
Classification: LCC HD61 .T48 2023 | DDC 338.5—dc23/eng/20221202
LC record available at https://lccn.loc.gov/2022056526

ISBN: 978-1-032-35892-5 (hbk)
ISBN: 978-1-032-35890-1 (pbk)
ISBN: 978-1-003-32922-0 (ebk)

DOI: 10.4324/9781003329220

Typeset in Bembo
by Apex CoVantage, LLC

Access the Support Material: www.routledge.com/9781032358901

Contents

Preface

Most of us have nearly zero control over the forces that impact us. We can't control the economy, the strategic initiatives of our organizations, or the decisions of others. Yet these forces impact us in all sorts of ways. Maybe the economy will reduce job opportunities, the new strategic initiatives will include the big project you have always dreamed of leading, or you will have one of the greatest ideas of your life while chatting with a stranger in a cafeteria lunch line.

In the risk science profession, we refer to this lack of knowledge as uncertainty. Regardless of how carefully we plan our futures, future projections are often wrong. When we manage risk, we consider how that uncertainty relates to important or severe consequences involving things we value. Risk and uncertainty are persistent forces that lead to problems, but could also introduce new opportunities that spur innovation.

Many academic fields attempt to understand risk and uncertainty. Statisticians use data to make inferences about larger populations. Forecasters seek patterns that can be used to predict the future. Financial analysts use probabilities to compute financial losses and returns. Many of these mathematical and financial advances have opened the door to incredible innovations in business and society.

Despite the massive datasets and complex mathematics, we as a society are constantly surprised. There are the black swan surprises, like 9/11. Some surprises seemed possible but were not sufficiently managed, like natural disasters and industrial accidents. There are surprises we let happen when we let our guard down.

A recent poll asked academics and students in the United States a simple question: What do you value in your life and career? These were their responses:

A good job, the ability to learn, good relationships, my health, my family's health, financial stability.

While money was a factor in these responses, there was still a clear overarching recognition that there is more to the story than just dollars.

The second question asked: When predicting the next ten years, what are your largest concerns?

My health, my job, the economy, climate change.

The third question asked: What can you do now to address those concerns?

Nothing, wait and see, get good grades.

Unfortunately, many of the respondents felt they had little to no control over their concerns, thereby suggesting we have little control over the future.

While nobody has a crystal ball to predict the world's biggest problems, there are tools and principles that allow all of us to have some control over *outcomes*. That is the concept of risk management. We "manage" risk to help influence the likelihood of risk events, the impact of those risk events, and the recovery after risk events happen.

This book provides a practical non-quantitative approach to understanding risk and making better decisions involving risk. Readers across industries and professions are eager to understand and manage risk. They are consistently tasked to manage activities that could lead to positive outcomes but are also acutely aware of the potential for negative outcomes.

You may be reading this book to learn how to understand and manage risk for many reasons. You may feel a sense of urgency to protect yourself, your communities, and your organizations from future risk events. You may feel overwhelmed by messages and polarization in the media, contributing to a sense of ambiguity over how to act in response to information or misinformation. You may feel intimidated by risk management topic areas because these topics are often portrayed as highly technical and probability-based, but seek to understand these topics by looking for the basic principles and not mathematics. You may seek to leverage an understanding of risk to manage risk in ways justified by evidence and reason. Most importantly, you may seek to use clear evidence and reason to find your balance between both positive and negative outcomes in an uncertain world.

Risk science researchers have made great strides in developing principles and practices for understanding and managing risk. These innovations, however, have historically been shared using academic textbooks and highly technical content. Currently, no existing work translates these risk science advances into practical and concise tools for nontechnical audiences. Using this book, you will gain an understanding of risk science and be able to apply that new knowledge in your professions and personal lives. While you cannot predict every possible risk event, you have control over how to address the risk, which could potentially impact the likelihood of risk events, the magnitude of those risk events, or the recovery after a risk event occurs.

This book will cover three main rules:

Rule 1: Risk is the single most constant and enduring factor in our lives and professions. Because we cannot predict the future with any amount of accuracy, it's expected that we disagree on topics of risk and uncertainty. That's not necessarily a bad thing. However, we can start to find a middle ground by recognizing that the amount of available knowledge can vary among situations. For example, our knowledge about fundamental laws of physics is pretty good, as we can make fairly good projections on the fate of an egg if it was dropped from a three-story building. The accuracy of our projections may be significantly worse if we were to predict the price of that egg 30 years from today. We also recognize that while uncertainty is unavoidable and risk management can help us, risk management cannot necessarily shield us from all negative events. However, we can instead think about how to treat the risk and how to recover from those risk events.

Rule 2: We do have some control, but not full control. When we accept this, we can actively manage that risk. While we can't shield ourselves from risk and uncertainty, there are a variety of practical steps we can use to manage risk. Contrary to widespread belief, risk management involves values and feelings and is not a purely objective construct. It's also important to understand the difference between risk and safety. We all have our own biases that impact how we think about and react to risk, and that's OK. We are also continuously influenced by value-laden messages in our social lives, media, and other interactions. That's OK too, but it's important to be aware of these types of forces.

Rule 3: Risk can be a good thing. Risk is not necessarily all about the bad events that can occur. Equally, good things can happen, and we can use the construct of risk to help balance between the good and the bad. There are many benefits of managing risk, which can help us be empowered to take control of the risk factors in our lives and professions. While uncertainty is pervasive, managing risk will always be worth the time and the energy.

This book will challenge how you understand and manage risk in your professions and personal lives. Don't worry about your memory of statistics. Risk is not just about probabilities and data. Even scientists with PhD degrees don't always agree on probabilities. There is always skepticism about data and information. Keep reading, and we will talk through these topics together.

The text of this book, including the online resource activities, will discuss many examples of highly visible risk issues. Some of these risk topics may seem alarming to some readers. Some of these risk topics are highly politicized, so individuals may already have stances on these issues. Some of these risk topics may have personally impacted readers and those in their communities. The goal of this book is not to share or promote the forming of political stances, particular perceptions, or reactions to the risk topics. Instead, the book aims to help think through these topics using a neutral third-party lens. We will further discuss those political and perception-related forces throughout this text.

Access the Support Material: www.routledge.com/9781032358925

Rule 1

Risk is the single most constant and enduring factor in our lives and professions

Rule 1 states that risk and uncertainty cannot be escaped and cannot be ignored. That is why the topics and discussion of this book are focused on understanding uncertainty and the knowledge involved with risk-related situations.

The general concepts of this rule refer to the activities studied during a risk assessment process. A risk assessment often consists of several components:

- Identifying potential risk events (e.g., asking "what can go wrong," "what can go right"). This consists of looking into history, asking experts, and leveraging data and models to identify potential risk events (positive and negative).
- Gauging the type and severity of consequences that could result from those risk events.
- Understanding uncertainty. The heart of risk involves acknowledging uncertainty when identifying potential risk events and gauging the consequences
- Evaluating the risk. This consists of forming some judgment about the risk and eventually determining whether a particular risk is significant, or acceptable/unacceptable.

The risk science approach involves conducting the discussed risk assessment activities using a variety of methods. For example, quantitative assessments can use probabilities and mathematical models. However, those activities also are supplemented with qualitative strength of knowledge judgments. These activities can also be fully qualitative, which is without numbers. In this book, we will keep the discussion non-quantitative, such that we are not focusing on math, but instead focusing on the concepts.

The discussion of this rule will happen over the course of three chapters:

Chapter 1 will discuss practical implications of risk science. At the basic level, the concepts behind risk science are simple and intuitive. What is new about risk science is how the conceptualization of risk has turned into processes and principles that can be applied to a variety of situations.

DOI: 10.4324/9781003329220-1

Chapter 2 will discuss main definitions used for risk, uncertainty, and related terms. You may be surprised to see that there are many differing and competing definitions of risk science. That is both a strength and an issue that is being solved as the risk science field grows. Multiple definitions can be a strength as it allows risk science to be specifically applicable to the varying practical uses, in which unique terminologies and best practices are deeply woven into the related standards and policies. But also, as the risk science field develops, additional commonalities among those varying definitions are being developed, and we expect to see agreement grow further in the upcoming years.

Chapter 3 will discuss the importance of considering available knowledge and the strength of that knowledge when understanding risk. This is in stark contrast to how risk is often portrayed in books and the media.

1 The heart of risk science makes sense

Learning Objectives

After reading this chapter, you will develop further knowledge on aspects of:

- Basic features of risk and risk science
- The importance of values in understanding and managing risk
- How to interpret risk in practical situations

Imagine it's the 1960s and the United States is in a space race to be a leader in the development of aerospace expertise, particularly human spaceflight. The newly formed National Aeronautics and Space Administration (NASA) is faced with the unprecedented task of sending an astronaut into Earth's orbit, and soon later, the moon. These initiatives involve incredible opportunity, enabling the United States to achieve one of its most ambitious goals, as John F. Kennedy declared, of "landing a man on the moon" and winning the arms race with the Soviet Union.

Despite this opportunity, there is also much that is unknown about the conditions of space travel, how engineering designs will perform in those conditions, and whether adequate safety plans are in place. There is potential for devasting loss of lives, property, defense capabilities, and political power.

The nation's leaders are called upon to manage risk. Without risk training, leaders may have no clear place to start. The problem is complicated by many factors. The political will toward achieving lunar mission goals is unwavering. However, space travel is a brand-new frontier, and engineering designs cannot be fully tested prior to launch. Without previous experience with space travel, it's difficult to even predict the issues they need to be concerned about.

Today, with vast amounts of research in risk science, we can begin to untangle the complicated risk issues involved. The concepts of risk science are used across many different types of applications, like those dealing with space flight, global pandemics, infrastructure management, public policy, cybersecurity, and many others.

DOI: 10.4324/9781003329220-2

Risk is about considering the future and asking how future situations or events impact those things we value. However, we know that our future predictions are almost always wrong. We call this factor uncertainty. In some cases, the uncertainty can be understood using probability numbers. In other cases, we may know so little about the situation, the most we can do is understand the uncertainty using words, beliefs, and opinions.

Another factor that makes space flight and really any other risk problem complicated is the issue of competing values. Before reading any further, grab some paper and pencil and make a list of things you value the most.

What did you write? Here are some common responses encountered over the authors' combination of 50 years of teaching risk to undergraduate students, graduate students, working professionals, and state/national agencies:

- My own health and safety
- The health and safety of my children
- The health and safety of my family and communities
- My finances
- Personal happiness
- World peace
- My house, car, and other property

Most respondents place high value on the well-being. Without that well-being, we can't effectively use our finances, cars, homes, and so on. To acknowledge this importance is a huge step in understanding the decisions we make to manage risk.

Here's the problem: These concepts of well-being are difficult to quantify. For example, apps that manage how we feel when we exercise have an interface that asks how we are feeling, as in Figure 1.1.

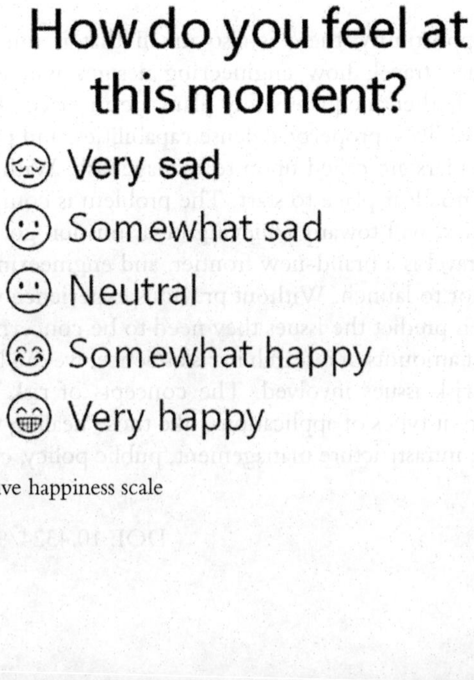

How do you feel at
this moment?

☺ Very sad

☺ Somewhat sad

☺ Neutral

☺ Somewhat happy

☺ Very happy

Figure 1.1 Qualitative happiness scale

How do you feel at this moment?

— 1 😔 Very sad

— 2

— 3 🙁 Somewhat sad

— 5 😐 Neutral

— 6

— 7 🙂 Somewhat happy

— 8

— 9 😁 Very happy

— 10

Figure 1.2 Qualitative happiness scale with numerical ratings

This is a very relative scale. How about instead attaching numbers to the scale shown in Figure 1.2?

Still, what's the difference between a 9 and a 10? How is my assessment of a 9 different from my friend's 9? Or the rest of the population? If I just heard my favorite song on the radio, am I more likely to rate myself as a 10? There seems to be some subjectivity with trying to measure things we often value the most.

One relatively easier measurement is dollars. These dollars serve as "units," allowing us to compare aspects of feelings, health, and environmental quality. However, imagine leaders during the space race converting all of their concerns to dollars, like the lives of astronauts, political power, and so on. It seems troublesome, but the reality is that there are many situations in which those types of conversions to dollars are relevant. However, the world is also changing and moving away from thinking about single financial metrics or single measures at a time.

Also, at the heart of understanding risk and uncertainty is the topic of knowledge. When we have little to no knowledge about these future predictions or the consequences of future risk events, this implies that we have large uncertainty. Science and other types of research attempt to gain the knowledge that can help us understand risk. We can then make our decisions by first asking ourselves how much knowledge we have and how well we trust that knowledge.

NASA recognized the low knowledge and extreme uncertainties related to the lunar space missions. They also understood the severity of the situation, recognizing the potential for disaster. This combination of uncertainty and potential for severe negative consequences led to NASA using a cautionary/precautionary stance. With this stance, they could have exercised caution by implementing a risk mitigation measure or deciding to not pursue the activity exposing them to the risk. For example, scientists were particularly concerned about lunar micro-organisms and their impact when astronauts returned to Earth. Due to the low knowledge involved with lunar micro-organisms and the many unanswered questions about space travel in general, NASA created the Mobile Quarantine Facility (MQF) used to house the returning astronauts and their scientific samples. This retrofitted Airstream camper would quarantine the returning astronauts for three weeks and serve as a major step in contamination prevention. While burdensome to the astronauts, this use of the MQF was a way to practice caution and to address the low-knowledge risk associated with lunar missions.

As a very different example, consider risk related to road travel. Imagine again, it's the 1960s. The National Highway Traffic Safety Administration (NHTSA)

Figure 1.3 NASA Mobile Quarantine Facility[1]

in the United States has just helped to promote the first seat belt law, requiring vehicle manufacturers to include seat belts in vehicles. While it took decades for additional regulation to require vehicle occupants to actually use those seat belts, the issue of seat belts and related risk is quickly gaining momentum, and quickly gaining knowledge. The function of a seat belt can be explained using principles of physics and biomechanics. The effectiveness of seat belts can be studied using available statistics through vehicle testing and accident data. These risk-related laws are supported using vast amounts of high-integrity knowledge.

Both use of MQF and the seat belt laws are also very different from the many other types of applications for one reason: These rules and laws involve human lives and health. One of the core values of risk science states that to protect human lives and health, the associated risk should be low.

The NASA and the NHTSA examples involve rules and policies. NASA required the use of the MQF. The NHTSA promoted inclusion of seat belts in vehicles. However, the responsibility for managing risk for the astronauts or vehicle passengers is much more distributed. For example, transportation safety risk responsibilities are spread across regulated vehicle manufacturing, technologies developed in private industries, and individual behaviors in which there are few rules or policies. While it's most effective for the responsibility for risk management to be allocated to those most equipped to control the risk, this in reality is impractical.

There is much more wiggle room in dealing with risk guidance when there are no rules or policies or when existing rules and policies are minimally enforced. We as a society understand the risks related to smoking and we see the labels on nicotine products, but the warning labels on those products have not eradicated smoking. One of the founders of risk science, Chauncey Starr, posed the question: "How safe is safe enough?" Because we all have differing values and circumstances, we have differing acceptable levels of risk. We also tend to be more accepting of risk when it is voluntary, as Starr said: "We are loath to let others do unto us what we happily do to ourselves."

All of the examples discussed in this chapter involved decisions related to risk-related policies, laws, or individual behaviors. It can sometimes be a long journey to go from understanding uncertainties and knowledge to the act of making a decision. All of these decisions are made using some level of management review and judgment in which individuals or groups consider the available knowledge, uncertainties, and any other information, and then use their own judgment to make decisions. Some of these decisions took decades in the making, while others are taken on a whim. Regardless, those tasked with risk science recognize that they can take ownership of the risk by systematically considering the issues and principles discussed here and in the upcoming chapters of this book.

Key Takeaways

- Future predictions are often wrong.
- What matters is understanding how knowledge impacts our under-standing of risk.
- Risk handling is about not only data/information/knowledge but also values.

Note

1 Photo from public domain, free of copyright restrictions https://airandspace.si.edu/collection-media/NASM-A19740677000-NASM2018-00640

2 We can agree to disagree on risk and uncertainty

Learning Objectives

After reading this chapter, you will develop further knowledge on aspects of:

* Basic definitions for risk concepts, including risk, uncertainty, and knowledge
* Reasons for disagreement on risk concepts

Let's first discuss some of the basic origins of risk. Risk as a concept has a long and complicated history. Some say concepts of evaluating risk for loans and business transactions for risk date back to 5000 BC. More than 2,400 years ago, the Athenians practiced assessment of risk before making decisions. That capacity was wholly non-quantitative, as these decisions did not use numbers, and consequently, they did not use concepts like odds and probabilities. Instead, they leveraged a more instinctive way of dealing with risk, by appealing to the gods and the fates.

Similarly, historians attest that risk understanding, risk management, and risk-related mathematics developed globally, ranging from ancient Egypt, Greece, India, and China. Some of the earliest documented scientific studies of risk were applied to games of chance and gambling. Many famous mathematicians of the 16th and 17th centuries applied probabilistic concepts to estimate a relative likelihood of some event occurring, often related to throwing dice. Blaise Pascal and Pierre de Fermat are well-known for their contributions to probability theory, which was the catalyst for groundbreaking advances that led to the birth of statistics, forecasting, and modern analytics. With these mathematical advances, we as a society could use past history, mathematical properties, and reasoning to predict future events, with improved levels of accuracy.

These probability concepts were particularly innovative and useful for the fields of finance and insurance, to understand financial losses and gains. In the early 20th century, there was a momentum toward applying probabilistic

DOI: 10.4324/9781003329220-3

and statistical concepts toward operational and management-related decisions. For example, in the early 1900s, William S. Gossett, an experimental brewer at a Guinness brewery, when tasked with increasing production quantity and decreasing costs, developed a major breakthrough in statistical reasoning, a reasoning that was the foundation of modern industrial quality control.

The industrial revolution of the late 1700s led to increased consideration of risk and safety. During this time, employment opportunities at new mills and factories led an unprecedented number of workers flocking to cities. Workers were faced with low pay, hazardous working conditions, child labor, and consequently, degraded safety and quality of life. Increased regulations continue to develop, as more attention is paid to risk activities worldwide.

The nuclear industry pioneered more advanced methods of quantitative risk in the late 1970s and the early 1980s. Due to the newness of the risk concepts and uncomfortable discussions about probabilities and risk judgments, these methods also met criticism. The WASH-1400, "The Reactor Safety Study" of 1975, prepared by the United States Nuclear Regulatory Commission is an example of this type of situation. The oil and gas industry is also known for pioneering more advanced risk methods beginning in the late 1970s and onward. This was largely in response to large-scale explosions, fires, and spills plaguing the industry. These new methods developed into commonly used tools for risk and safety assessments that are still used today.

Also in the early 1980s, the *Risk Analysis* journal was created to formally introduce risk as a new discipline. The inaugural issues of the journal discussed topics including quantitative definitions of risk, risk in energy production, risk of carcinogens, occupational asbestos exposure, and cost-benefit analysis.

Risk science, especially when directed toward security applications, gained momentum following the September 11, 2001, terrorist attacks. This event prompted nations, communities, businesses, and individuals to take more concerted efforts toward addressing risk and safeguarding things we value. As a result, academics and professionals across the world, and across academic fields, have focused attention on developing methods, approaches, and best-practices for understanding risk and managing that risk.

Risk and safety remain a worldwide issue, as risk concerns (e.g., health, worker safety, food availability/safety, nutrition, child/forced labor, environmental issues, terrorism, and too many others) exist. As a result, an increased understanding and dedication to addressing risk remains incredibly important for us all.

While risk concepts and methods continue to develop, we also recognize that individuals maintain an intuitive understanding of risk concepts. This intuitive understanding could include elements of a danger, potential undesirable consequences, or losses. When risk professionals attempt to formally conceptualize and measure risk, they consider more refined definitions and metrics. However, risk science shows how intuition and formal theories about risk can live side by side. The key is to distinguish between the concept of risk as reflected by the intuitive understanding and how to describe it or measure it in the risk assessments. There are multiple ways of conducting the measurements; the way that is the most suitable will follow general risk science guidance tailored to the situation. The risk assessments build on knowledge and judgments which are not objective. Hence, the risk characterizations and how the risk is understood are subject to discussion. There could be different views. When considering how to handle the risk, the difference could be stronger and more visible, as the handling is not only about knowledge but also values.

Let's clarify an essential point: Risk science includes judgments. While we prefer to be informed by the most trustworthy data, scientific research, and information, we have to eventually form judgments to characterize and manage risk.

Suppose you and your friend have decided to begin a fun and fulfilling new hobby. You have done your research and have found that hiking in national parks is a perfect choice. Fresh air, exercise, and time away from work and electronic devices could add a new and exciting dimension to your life. You even see the potential for other outdoor hobbies once you develop the strength and agility of a hiker.

However, your friend disagrees. He heard a story about a friend-of-a-friend's hiking incident that happened a few years ago. He would prefer a safer alternative. While no activity is risk-free, your friend has deemed hiking to be an unsafe activity with an unacceptable risk.

Neither of you is wrong. You just have different characterizations of risk. Let's work through your characterizations. This characterization contains three elements: The consequences, uncertainty judgments, and background knowledge. Consider these three elements to be different lenses with which to view the risk. There is considerable overlap among these three elements.

You and your friend have a massive disagreement over the risks of this new hobby. Your friend is convinced there is going to be a severe negative consequence based on the evidence from the story he heard. He gives weight to this story for all kinds of reasons, some of which relate to feelings and perception and others relating to the fact that he has never hiked in a national park before, and this story constitutes a major portion of his experience with the topic. You consider his supporting knowledge to be very weak because it's based on a single story about an event that seems rare and preventable with the appropriate precautions (precautions you would take), but he has high trust in the information source and his own judgment.

You see the potential for positive consequences. You find some statistics about hiking safety. You have several conclusions:

- Fatalities in national parks are rare, citing that the likelihood of dying in a national park is nearly zero.
- Many causes of the deaths that occurred in national parks were common causes of deaths outside of the national parks, such as those related to vehicles and water.

You also sense certainty using the information you have collected. Your knowledge is strong because you have read multiple reports and news articles supporting your stance.

However, your friend still disagrees.

You both have used your available evidence to form your judgments about risk: Addressing the consequences, likelihood, and background knowledge. You also have different risk perceptions that influence your stances. You can agree to disagree on these issues. However, risk science can also offer guidance on thinking through this disagreement.

You and your friend are focused on different aspects of the risk. You concentrate on the positive benefits of this activity, along with the dimensions of health and well-being. Your friend focuses on the possibility of a severe accident or death. Because you both emphasize different core values, there is no right or wrong value to consider.

You and your friend also place different weights on the uncertainties involved with the activity. You are relatively certain that this activity would result in some typical or expected outcome. Your friend focuses on many uncertainties, possibly the weather, his own physical ability to perform the tasks, and the existence of dangerous wildlife on the trails.

You and your friend also leverage very different knowledge. Your friend relies on the memory of a single incident. You also found information that included analyses using data on past incidents. Despite the differences in the information sources, you both consider your strength of knowledge is sufficient to make a decision. However, an objective observer may not agree that knowledge integrity is sufficient for either of you.

You and your friend also have a differing sense of vulnerability. Vulnerability represents the potential for some undesirable consequence, given an event. Based on the uncertainties and knowledge, your friend senses a higher potential for serious consequences in case a risk event occurs, and consequently a higher level of vulnerability.

You have a higher sense of resilience compared to your friend. Resilience refers to the ability to return to normal after the occurrence of a risk event. Your friend may sense that an incident with this new hobby could lead to a condition that is permanent or would result in a long recovery. You instead focus on less serious incidents that you could recover from relatively quickly, with no permanent losses.

As a result of the differences in interpreting risk issues, you and your friend have a differing sense of *safety* for this new hobby. While many define *safety* as the absence of accidents and losses, this definition doesn't incorporate the uncertainty component that we consider with risk. Instead, we define safety as an acceptable or tolerable risk. If the risk is low, we say the safety level is high, and vice versa.

This sense of safety is based on our judgment of the risk. You judge the risk of this new hobby to be low and therefore judge the risk to be acceptable or tolerable, and consequently consider this hobby to be safe. We can generalize that safety encompasses an acceptable or tolerable level of risk when considering all possible risk events associated with this hobby, such as accidents, thefts, and poor weather conditions. This is in contrast to *security*, which generally refers to an acceptable or tolerable level of risk when considering only malicious risk events, like those that could result from an attack or theft.

Differences in risk characterizations are widespread across personal and professional settings. Let's say you work for an energy infrastructure provider. Your company generates electricity using methods like nuclear, natural gas, wind, solar, coal, and hydroelectric. That energy is critical for charging our cellphones, functioning cellular networks, hospital intensive care units (ICUs), manufacturing life-saving medicines, powering electric cars, and so many other elements of our society.

As a result of the many different uses of the energy infrastructure, various stakeholders evaluate consequences, vulnerability, resilience, and other risk-aspects of infrastructure in different ways. For example, the various stakeholder viewpoints could be:

- Customers: Want the infrastructure to be fully functional at all times; they also want the rates to be affordable
- Other dependent infrastructure providers (e.g., communications and healthcare): want the infrastructure to be fully functional at all times
- Shareholders (if any): Want an increased stock value and overall maximized value of the corporation
- Environmental groups: Want the energy company's operations to reduce any negative impact on wildlife, air quality, water quality, and so on.
- Community members around energy infrastructure facilities: Want the facilities to not impact community quality of life (e.g., water quality, air quality, noise, and traffic)
- Employees of the infrastructure provider: Want steady work and compensation

These stakeholders all have varying considerations. Some stakeholders, such as customers and shareholders, have near-term objectives. Customers want working infrastructure today and shareholders want to see returns today. Other stakeholders may be more concerned about the longer term. Environmental groups concerned with climate change are concerned about a much longer time horizon.

Some stakeholders may be more concerned about as-is or usual operations versus those in cases of emergencies or risk events. For example, community members may care a lot about restoring power in cases of a major storm. Dependent infrastructure providers may also have their own risk plans around the possible loss of energy. These dependent infrastructure providers may invest in generators or backup-energy methods to maintain their critical operations.

Some stakeholders may be unaware of the interdependencies impacting their various concerns. A scenario that is widely discussed in the risk literature involves interdependencies among various infrastructure types. In this scenario, an energy infrastructure facility loses functionality. As a result, the dependent communication infrastructure also loses functionality. As a result, there could be potential for additional energy infrastructure facilities to lose functionality. This cascades into a major blackout that disrupts the entire community, from healthcare and transportation to manufacturing. There could also be global implications as the affected area could produce critical medicines, house cyber-resources necessary for global cyber-operations, be a hub for transportation networks, and so on.

As a result, the issue of consequences of a loss of functionality for this energy infrastructure is much more polarizing in this setting compared to the hiking example. Those consequences can impact people, businesses, and the environment in many different ways. We can generalize that people's safety and well-being are impacted by power loss when considering functionality of hospital ICUs, road safety, and manufacturing of life-saving medicines. The consequences also include dimensions of fairness and equity when asking how an electricity loss impacts different populations in different ways.

The issue of uncertainties is also equally polarizing. Energy providers are concerned about risk related to the loss of functionality of their system, leading to disastrous consequences. That loss of functionality could result from severe weather events (e.g., hurricanes and winter storms), earthquakes, terrorist attacks, and solar storms. Experts think these events can occur, but nobody can be sure if these events will happen in the next year or decade. Hundred-year floods are a classic example. Statistically, these floods happen every 100 years. However, the world is not static, so the likelihood of these floods may be continuously changing. If the infrastructure provider spends millions protecting against this flood through increased rates to customers who may struggle to pay for services, and if the flood doesn't happen, the result could be a public relations nightmare. If this flood happens and the infrastructure is not sufficiently protected, the resulting aftermath could end in a humanitarian disaster.

The issue of knowledge is also an enduring issue. While our high school textbooks teach us that science concepts are fact, the boundaries between fact, theory, speculation, and misinformation are often up for debate when working with real risk applications. Across the world, experts analyze weather and climate data to develop predictions for future weather events and climate conditions. These experts often disagree. Consider a hurricane spaghetti (ensemble) model, as shown in Figure 2.1. News sources share maps showing the potential hurricane tracks when climate conditions are ripe for a hurricane. Some paths may lead to our own towns, while others steer clear of us. These hurricane paths are different models created using highly qualified scientists using expensive and incredibly advanced computing resources. The predictions have some degree of error, which can be expected when studying such complex global systems, despite the time, cost, and energy used to create these models. As another example, consider the long-range predictions for climate conditions we often refer to as "global warming." Scientists and non-scientists

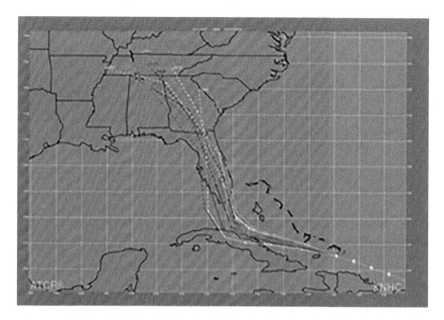

Figure 2.1 Ensemble models showing many different paths for a hurricane[1]

alike vehemently disagree on this topic. The fact that the subject is embedded in political agendas adds fuel to the discord.

It's also difficult to find agreement on risk and uncertainty because of the many different accepted practices. Even some commonly used practices are flawed from a risk science perspective.

Let's consider the management practice of prioritizing risk. Let's say the infrastructure company mentioned in the previous example seeks to identify the most important risk concerns it should consider addressing immediately. It's a common practice to identify the top 10 or the top 20 risks needing further study. The selection of 10 or 20 risks seems arbitrary and should be avoided. What if more risks were particularly pressing?

As another example, consider the "risk matrix." A quick web search on the term found about 955,000 search results, so it's a popular approach. However, it's also profoundly flawed. This risk matrix is a way for risk analysts to identify the top priority risks. Each risk event is assigned a *likelihood* and an *impact*. The *likelihood* is like a probability of the event occurring. The *impact* is like a consequence. We already talked about how consequences are difficult to predict and cover many dimensions. Probabilities are equally troublesome. Even PhD scientists struggle with explaining their meaning. According to these risk matrices, risks in the top right corner (high *likelihood* and high *impact*) have a relatively higher *priority*. These risk matrices don't acknowledge differences in supporting knowledge. What if a risk shows a *high impact* and a *low likelihood*,

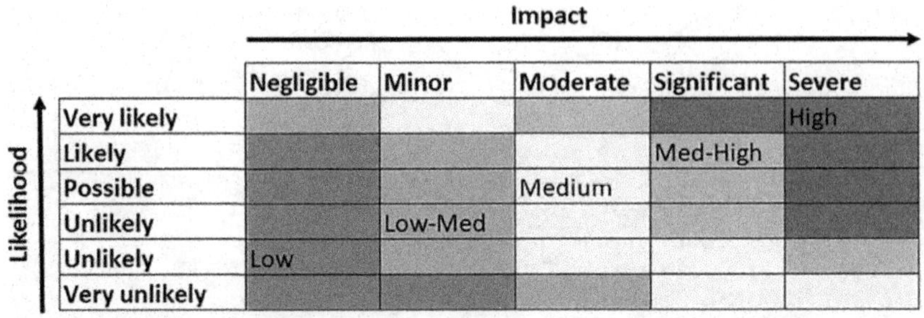
Impact
Negligible
Minor
Moderate
Significant
Severe
Very likely
Likely
Possible
Unlikely
Unlikely
Very unlikely
Likelihood
High
Med-High
Medium
Low-Med
Low

Figure 2.2 Example of a risk matrix

yet very little is known about the phenomenon? For example, this event could involve a company considering the loss of functionality after a cyber-attack on their infrastructure. Should this risk have a high priority? According to this risk matrix, the event might fall into the medium priority. However, the selection of priority level is really a judgment that shouldn't be trusted to a simple 2-D matrix, as shown in Figure 2.2.

There are also many sources for standardized risk management practices, through sources like Enterprise Risk Management or the ISO 31000 framework. We won't go into the details of these differences, but these standards are noticeably different from one another in their definitions of risk and in the processes involved.

There is also a long history of asking general questions to understand and characterize risk. A well-known method still very commonly used today is Quantitative Risk Assessment (QRA). This method is based on the triplet developed by Kaplan and Garrick more than 40 years ago:

- What can happen (i.e., what can go wrong)?
- If it does happen, what are the consequences?
- How likely is it that these events/scenarios will occur?

While the conceptual questions in this triplet need to be supplemented by aspects of knowledge and other concepts presented in this book, it provides an essential structure for understanding and characterizing risk.

The answers to these questions can differ from person to person. Even analysts and experts may form different opinions on these questions. When these questions are used correctly and in situations where experts can credibly answer these questions, we may see some consistency among experts. A lack of consistency among individuals, analysts, and experts answering these questions could suggest large uncertainties or different knowledge of the systems and activities studied among those persons.

While the answers to the triplet questions can be very useful, there is a need for expertise and a basis for the answers. Otherwise, we make the answers appear credible and trustworthy when they have a poor foundation that lacks real credibility. Consequently, we end up misleading ourselves and the audience of the risk study.

Different fields and professions also think about risk very differently. The finance field often studies risk by thinking about factors such as variability in financial instruments, existing or changing regulations, and the preferences of investors. Because there is a common use of monetary units and standardization in these calculations, it becomes easier to find some common ground when discussing risk assessments among financial analysts with comparable training. In a healthcare setting, risk issues involve patient safety, regulations, malpractice, and existing or emerging policies. These issues cut across many dimensions in ways that are not as easily placed into calculations, making consistency even among those with similar training challenging. Yet also the health field has some typical forms of conceptualizing and describing risk. In an operations setting, risk issues often cut across dimensions, involving issues like the protection of the functionality of the business, the safety of people, or the quality of a product.

It's reasonable and expected that different fields confront risk differently. However, the best practices and main issues of risk science are firm across dimensions, which we will discuss in the upcoming chapters.

Key Takeaways

Different fields and applications have varying definitions of risk concepts.

- The best practices and main issues of risk science are firm across dimensions, which we will talk about in the upcoming chapters.
- Safety refers to acceptable or tolerable risks.
- Issues of risk-related definitions can be polarizing in both political and societal contexts.

Note

1 Sourced from government website www.weather.gov/news/181406-director-cautions, not copyrighted www.omao.noaa.gov/find/media/images/image-licensing-usage-info

3 Risk is about knowledge

Learning Objectives

After reading this chapter, you will develop further knowledge on aspects of:

- The use of knowledge in understanding and characterizing risk
- Knowledge as it relates to highly visible risk-related terms, like "black swans" and "perfect storms"

Take a moment and think about all of the surprising world events that have happened in recent history.

Maybe your list included:

- 9/11
- 2008 financial crisis
- Fukushima nuclear disaster
- COVID-19 pandemic
- Russia–Ukraine War

Maybe your list included events more specific to your personal and professional life. When the authors poll professionals about the most surprising and impactful events in their lives, answers include:

- Accidents
- Natural disasters
- Chance meetings that turned into future friendships
- New job opportunities

Even if your list is different, ask yourself: Before the event occurred, did you really think the event would occur? Or would you have been able to use a probability number to describe the likelihood of that event? A common answer

DOI: 10.4324/9781003329220-4

is *no* to each of those questions. Maybe we knew an event was *possible*, but we may not have taken the event seriously enough to act on that possibility.

One way to reframe our understanding of those events is to examine our level of information and strength of knowledge of those types of events before they occurred. When we talk about those types of events, we think in general terms. For example, when we talk about natural disasters, the kind of event could be a "hurricane." We are not specific about the date of the hurricane, the impact, or the magnitude.

In 2002, the U.S. Secretary of Defense, Donald Rumsfeld, famously used the "unknown known" term as "the things that you think you know that it turns out you did not" when discussing weapons of mass destruction in Iraq. While he was not the first to use the term, he did effectively publicly introduce the concept of general knowledge/awareness versus that of a specific group.

Figure 3.1 allows us to frame that categorization of knowledge/understanding. The horizontal axis shows general understanding/knowledge about a particular risk issue. The vertical axis shows a specific group's general knowledge/understanding.

Let's say we are A, the risk assessor, and B is another entity, for example, the scientific community.

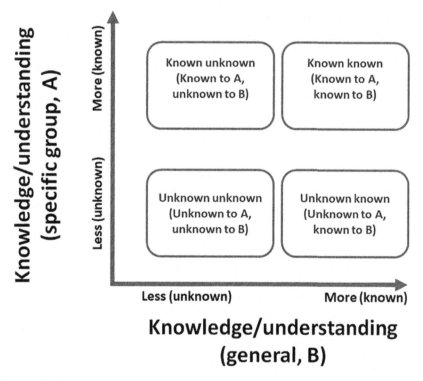

Figure 3.1 Classification of knowledge/understanding for the risk assessor (A) and some other entity (B)

Unknown knowns refer to things that are unknown or not understood by us (A), but are known/understood by others (B). This could involve national defense issues, as in the case of Donald Rumsfeld's famous quote, or it could involve lesser-known personal events. Suppose you're buying a house in a hot home-selling market. You have a little credible knowledge about the outcome of your particular bid on your dream home. Still, the realtors and homeowners have the knowledge about other offers and conditions to accept a specific bid.

Unknown unknowns are situations that are not known/understood by both A and B. These are the unforeseen events we would not have been able to predict. For example, during the early 1980s, in the early history of the AIDS pandemic, the occurrence and spread of the HIV virus and AIDS disease were unknown. As a second example, the drug thalidomide was used in the 1950s as a treatment for various health conditions, including symptoms of early pregnancy. Despite being approved for marketing and distribution in several countries, the use of the drug resulted in tragic outcomes. Children born to mothers who used the drug were observed with gross limb malformations. Practices for testing the safety and efficacy of drugs were not as developed compared to today (though today's standards are also constantly evolving), leading to these health effects to be unknown.

Known unknowns are situations that are known/understood by A, but not known/understood by B. This might serve as an advantage to A, if A is leveraging opportunities. Let's say A has developed a method to grow the best smelling and most beautiful flowers. Then A can use this as an advantage for expanding their flower business. When considering downside risk, this could be a situation in which A is an attacker or a risk source.

Known knowns are situations that are known/understood by both A and B. Suppose A manages a factory that produces car bumpers. The manufacturing line is highly engineered and monitored. A knows about as much as anyone can know about the status and quality of the finished products. Surprising events, like black swans (to be discussed later in this chapter), can still occur.

Using this type of classification, we can begin to reinterpret those events and see similarities among them, as in Figure 3.2. For example, those chance encounters and coincidences that led to new jobs and friendships fall in the lower left quadrant of the chart, signaling that these are generally *unknown unknowns*. Conversely, issues of national security and terrorism are in the lower right quadrant, as they are *unknown knowns*. These issues may be poorly understood by us (entity A), while potential attackers (entity B) have much more knowledge/understanding. Events like natural disasters and financial crises could be placed toward the lower center of the chart. We (entity A) may have some working knowledge of the issues but also understand that experts (entity B) have much more knowledge.

We could then use this classification to see what new information we should seek to better understand the most important risks. For example, for the most important risks, we can identify ways to move higher on the vertical axis. We

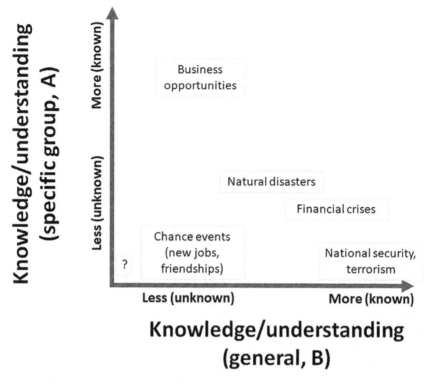

Figure 3.2 Examples of risk events classified using knowledge/understanding for the risk assessor (A) and some other entity (B)

can invest more time and energy into seeking new friendships or learning how to identify the potential for natural disasters. Suppose we are in positions of power, such as in defense, or are scientists. In that case, we can also work to move high-priority risks further to the right on the horizontal axis concurrently with moving upward on the vertical axis.

The incidence of 9/11 is an example of a black swan event. Black swans have been described in the literature for centuries, as it was commonly used in 16th-century London to describe something that is thought to be impossible. The concept of the black swan event was made popular by Nassim Taleb, who further describes the origin of the term:

> Before the discovery of Australia people in the Old World were convinced that all swans were white, an unassailable belief as it seemed completely confirmed by empirical evidence.

> (Taleb 2007)

Black swans actually existed in Western Australia but were yet to be discovered by a particular entity. In 1697, a Dutch expedition to Western Australia, led by Willem de Vlamingh, discovered black swans on the Swan River. The black swan was a surprise for "us" relative to our knowledge (not for them the inhabitants of Western Australia). The black swan came to represent something extremely rare, perceived to be impossible, but is later disproven. If we have no awareness of a particular event, condition, or fact, we're inclined to think that it's impossible.

Taleb (2007) further defines these events as:

- An outlier, as it is beyond our regular expectations (past history makes it not seem possible)
- Has a large/extreme impact
- Despite being an outlier, we develop explanations for its occurrence after the fact (i.e., we believe the event was explainable and predictable).

The basic idea is that a black swan is a surprising/unforeseen extreme event relative to one's beliefs/knowledge. They can occur for any of the known/unknown combinations.

The financial crisis in 2008 can also be viewed as a "perfect storm," which is a rare confluence of well-known phenomena creating an amplifying interplay leading to an extreme event. While the term is often used to refer to meteorological events, it can apply to any type of risk event. These perfect storms can also be viewed as a black swan. For example, suppose you found a new and exciting job opportunity after a set of coincidences. You met an industry executive in line at a restaurant. That executive mentioned being late for a meeting related to a new project, an unexpected job opening led to you being assigned to a related project in a different industry, and you both found a perfect match. This new job changed your entire career trajectory.

The perfect storm concept is also related to the concepts of dependency and interdependency. Interdependent systems are connected such that some event or change in one part of the system may influence other parts of the system, and vice versa. In other words, one risk event, even if seemingly small, can potentially influence another risk event. There is the well-known concept of the butterfly effect. A butterfly flapping its wings in one part of the world can impact the global weather. This concept is commonly studied from a risk perspective. Examples include:

- Infrastructure: Understanding how disruptions in infrastructure impact related infrastructure and other dependent entities (e.g., how disruptions to energy infrastructure can impact communications, and potentially how those disruptions to communications can cause additional disruptions to energy).
- Medicine: Understanding how problems with an aspect of health can impact related aspects (e.g., how issues with the respiratory system can relate to the cardiovascular system).

- Manufacturing: A shortage or a delay of a single component can impact the ability to produce a wide variety of other products and components.
- Transportation: Congestion or delay in one part of a transportation network can impact the performance of transportation across the entire network.
- Food production: Bees pollinate crops needed for human and livestock consumption. Pesticides, such as insecticides, fungicides, and herbicides, can disrupt the availability of bees (e.g., kill bees and impair reproduction). These issues can further lead to changes (possibly increased dependence) on the use of pesticides.

These concepts of dependency and interdependency have helped researchers and individuals more carefully see their risk problems as a system of interconnected parts, or systems of systems. In other words, we recognize that no entity (individual, group, object, etc.) stands alone in isolation but instead is impacted by others and has an impact on others. Thus, risk events, whether major or minor, can propagate through the system, causing effects on other parts.

You might be wondering why all of these concepts and definitions matter for the purpose of risk science. One key reason is that conversations about these metaphors help us better understand risk and encourage us to think more carefully about surprises related to risk. Risk events can be complex and are understood using many different perspectives (black swan and perfect storm) to describe a single phenomenon (rare, surprising, and extreme events). Comparing these perspectives can help us interpret historical risk events and also help us make risk-based decisions when considering future risk.

While the risk terms we have discussed in this chapter have been groundbreaking in the business and risk professions, we're also careful because these types of metaphors can also highlight and hide important facets of a risk problem. If we overly focus on one or few aspects of the risk, we lose sight of the big picture (as we cover throughout this book), including the many other aspects of the risk problem.

Bringing attention to the inability of risk assessments and risk science to foresee risk events could point to shortcomings in the approaches and methods used. However, the prediction of events is not the sole intent of risk assessment and management. Proper implementation of risk assessments and management also includes other activities, like:

- Keeping a healthy and well-managed system that can address new risks when they emerge
- Maintaining an agile system that can change course or quickly react when conditions change
- Maintaining a resilient system that can bounce back quickly and effectively when risk events occur (whether or not those risk events were predicted).

More recent advances in risk science also make the distinction between low-knowledge and high-knowledge events. Suppose you are planning a summer picnic and are concerned about risk related to rain and storms on the day of the event. You have weather forecasts to help you decide whether to cancel the event, experience with shelters at the park, and a lot of past experience dealing with these storms if they were to occur on the day of the picnic. Picnic planning is a high-knowledge situation, which you could treat very differently compared to a low-knowledge situation. Suppose, instead, you are attending a potluck event in which many strangers bring food to share. You know very little about the preparers of the food, how the food is prepared, and how long the food sits out in the sun. You have serious concerns about the safety of the food in this low-knowledge situation. You may exercise more caution in the low-knowledge situation versus the high-knowledge situation.

The discussion of terms and the distinction between low-knowledge versus high-knowledge enable us to think carefully about how we assess the risk. The next chapter will more carefully talk about how we can manage risk.

Reference

Taleb, N. N. (2007). *The black swan: The impact of the highly improbable* (Vol. 2). New York: Random house.

Key Takeaways

- Low-knowledge versus high-knowledge can be a critical factor in risk-based decision-making.
- A black swan is a surprising/unforeseen extreme event relative to one's beliefs/knowledge.
- A perfect storm is a rare confluence of well-known phenomena creating an amplifying interplay leading to an extreme event.

Rule 2

We do have some control, but not full control. When we accept this, we can actively manage that risk

Rule 2 demonstrates that all individuals and organizations have some control over managing risk. The discussions of this chapter include some main concepts of risk management, risk perception, and risk communication.

Rule 2 recognizes that we have several options when considering how to address some known or unknown risk. We deliberately choose among those options by not only reflecting on the available knowledge, data, and information but also reflecting on our individual or group values. Our stances on those risk topics might also be influenced by biases and subtle cues in risk messages. By understanding our own biases and propensity to be influenced by risk-related messaging, we can better enable ourselves to take greater ownership of our risk-related decisions.

Chapter 4 will discuss some best practices for organizations that are known for exceptional risk management capabilities. The chapter will also discuss the main options we have in addressing a particular risk.

Chapter 5 will discuss the importance of understanding the integrity of information and knowledge when evaluating risk. We distinguish between misinformation and disinformation and relate those concepts to general characteristics of risk science. Because risk involves concepts of uncertainty and future predictions, that uncertainty can be exploited to form misinformation and disinformation.

Chapter 6 will discuss cognitive biases that can impact how we perceive and react to risk events. Many of these biases are so familiar and ingrained in our thought processes, we may not be consciously aware of them. However, it's essential to recognize these biases and understand how they may influence your risk-based decision-making.

Chapter 7 will discuss basic concepts of risk communication. The concepts apply to stakeholder engagement, communicating with others, and interpreting communication presented to us. We are also mindful that some communications may have the intent to mislead, requiring us to be extra vigilant.

DOI: 10.4324/9781003329220-5

Chapter 8 will discuss various non-financial metrics that can be important to consider within risk science. More and more so, these nonfinancial metrics are changing the way agencies and corporations measure performance. There is also discussion about ethics as it relates to risk, as issues of fairness and justice in risk and risk outcomes are a serious societal issue.

4 You have some control

Learning Objectives

After reading this chapter, you will develop further knowledge on aspects of:

- The options individuals and organizations have to manage risk
- The importance of developing individual and organizational resilience for addressing risk

Think back to your experiences witnessing or contributing to an impeccably managed team. Maybe it was a silent restaurant kitchen where everyone focused on their tasks. Maybe it was a surgery in which the surgeons and medical team were highly trained for very specific operational roles. It could have been a firefighting operation or your favorite soccer team. The members of these operations remain calm and coordinated with others despite the chaos surrounding them.

Those suggestions in your high school yearbook were true. There is value in keeping it cool, especially in the most chaotic situations. The most emotionally intelligent professionals develop skills for keeping it cool in many ways, including maintaining emotional self-control, awareness of one's own emotions, being socially aware, managing conflicts, and working in teams.

These qualities relate to individual people, teams, and entire organizations. While each quality shown earlier is essential, it's most beneficial to complement those qualities with situational awareness and to maintain control.

At the organizational level, organizations that maintain control through performing complicated and sometimes dangerous tasks with a low error are known as high-reliability organizations (HRO), a concept developed by Weick and Sutcliffe, and others. These organizations can be characterized by using several qualities:

- Preoccupation with failure. The team focuses on failures or near misses, learns from them, and works to prevent future occurrences. This is

DOI: 10.4324/9781003329220-6

a dramatic culture shift for many organizations. It's human to not like criticism or to resist having their failures be put in the public spotlight. It's an uncomfortable yet critical task for organizational growth and management.

- Reluctance to simplify. The team is aware of the highly complex conditions involved with their tasks. Their own experiences provide only a single lens of a situation, and that perspective needs to be supplemented by other viewpoints. Seeking input from diverse perspectives requires a diversity of experiences and opinions. Imagine a boardroom with high-ranking executives across facets of a company. This board could be responsible for seeking input from professionals and making risk and safety decisions for everyone on the company's payroll, including those who work in a climate-controlled office building, those who work in refrigerated warehouses, those who work in 100-degree Fahrenheit conditions on oil rigs, and anything in between. Some stakeholders may be largely unaware of the "whys" of risk and safety-related metrics, and there may be little to no insight from the majority of stakeholders who are most impacted by risk and safety-related decisions.

- Sensitivity to operations. The team is responsible for monitoring current conditions and is sensitive to changes. It is imperative for team members to speak up if they notice any changes to current conditions. This is an act of situational awareness.

- Commitment to resilience. The team recognizes that surprising and unforeseen events may happen, regardless of whether the event is caused by errors, changes in conditions, or anything else. The team must be prepared to detect errors, maintain functionality, and recover when the surprises happen. Suppose you work for a state transportation agency. Their job is to ensure that the roads across the state are passable. If the agency doesn't fulfill that mission, it could disrupt the movement of ambulances, fire trucks, school buses, and the movement of goods and services across the state. Yet things go wrong all the time. Accidents, snow storms, hurricanes, sinkholes, maintenance, and a whole slew of other events. Resilience is imperative in these situations.

- Deference to expertise. Regardless of rank or hierarchy, when situations arise, decisions are made by the team member with the expertise to solve the problem. This contradicts traditional management structures in which the highest-ranked entity is responsible for making decisions. This also calls to question organizational models in which corporate boards make risk and safety decisions for all levels of an organization.

The practices of the HRO team members can provide useful guidance for the busiest managers and executives. Busy professionals often use the phrase "putting out fires" to represent the constant pressure to handle unexpected and crucial tasks. The time spent putting out fires could be spent on developing strategies to avoid those "fires" and making long-term planning decisions. Of

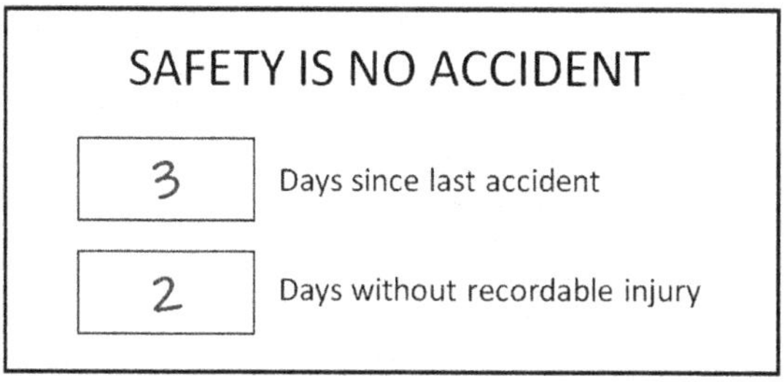

```
┌─────────────────────────────────────────────┐
│                                               │
│         SAFETY IS NO ACCIDENT                 │
│                                               │
│     ┌──────────┐                              │
│     │    3     │    Days since last accident  │
│     └──────────┘                              │
│     ┌──────────┐                              │
│     │    2     │    Days without recordable injury │
│     └──────────┘                              │
│                                               │
└─────────────────────────────────────────────┘
```

Figure 4.1 Simple example of metrics that matter

course, finding that time to maintain control will always be a challenge. Taking steps early to maintain control is particularly rewarding in risk-related situations.

Following the HRO's characteristic of situational awareness, we follow the advice of Peter Drucker, who famously said, "what gets measured gets done." While the origins of the phrase have also been debated, this quote alludes to the need to continuously monitor operations. Not everything important can be measured. Companies often create dashboards that can help monitor situations, a very simple but effective reminder as given in Figure 4.1.

While dashboards are very effective for maintaining control and situational awareness, there is always concern over unintended consequences. Suppose the safety metric used for the sign in Figure 4.1 was linked to salaries or bonuses for employees:

If an accident occurs, no bonuses for the current month.

There are very serious unintended consequences involved with the safety metric policy as shown earlier. Suppose an employee slipped on spilled oil in a vehicle maintenance facility. The employee was a little bruised but otherwise OK. The coworkers encouraged the hurt employee not to report the accident because reporting the accident would lead to a loss of a monthly bonus. The employee didn't report the accident. An hour later, another employee slipped on the same oil spill and wound up in the emergency room. In this example and others, it may not always be clear whether the consequences associated with this safety metric policy are intended or unintended. Some organizations with poor risk stances may use these types of safety metric policies to deter the reporting of accidents in unsafe working environments.

The use of dashboards and conversations about risk create a risk culture. There are also more direct ways to maintain control in the face of potential risk events. Let's say your family has happily welcomed a new addition: A sweet and beautiful six-year-old dog. You soon realize this 75-pound puppy doesn't fit well in your small car. You go shopping for a larger vehicle. You have done your homework and identified a few used vehicle models that you foresee as an excellent fit for your lifestyle and family. Here's the problem: A few weeks ago, a strong storm barreled through your region, leaving major flooding in the aftermath. You're very worried about the risk related to potentially purchasing a flood-damaged vehicle. While a title check could reveal flood damage, you have heard some stories about customers inadvertently purchasing flood-damaged cars, even when working with reputable dealerships. You have identified the possibility of buying a flood-damaged car as a risk with four potential options:

- *Accept the risk:* You're OK with the possibility of inadvertently purchasing a flood-damaged car, particularly if the damage from the flooding is minor anyway.
- *Transfer the risk:* You're willing to pay extra for a certified pre-owned vehicle. If there are flood-related or even non–flood-related issues with the car later, they would be covered by the included warranty.
- *Mitigate the risk:* You could take action to reduce the risk related the purchasing a flood-damaged car. You could do your own title searches before thinking seriously about a car. You could educate yourself on identifying the signs of a flood-damaged car. You could choose to only shop for cars through reputable sellers, even if those cars are sold at premium prices.
- *Avoid the risk:* You could decide not to purchase a car yet. Or you could instead buy a new car. You could wait a few months to avoid the possibility of encountering flood-damaged cars from this particular storm. Or you could search for cars in other regions that were not impacted by this storm or similar storms, though recognizing that car inventories travel across the country.

The four options mentioned are commonly known as the four primary risk management options. None of those options are necessarily best across the board. Instead, they depend on the decision-maker's situation and attitude toward the risk. You may need the car right away to be able to transport the dog to veterinarian appointments, so you may choose to accept the risk. You may foresee yourself losing sleep over the potential for a mechanical failure after accidentally buying a damaged car, so you may choose to avoid the risk by purchasing a new car. You may instead feel a sense of control by choosing to mitigate the risk. Learning about how to identify damaged cars puts you in a position of power and allows you to play a more active role in managing the risk for yourself and your family.

The used car example involves a buyer who has already identified a risk of concern. What if you're worried about all the potential problems with purchasing a car. The unexpected could include:

- Flood damage
- Buying a vehicle that doesn't fit your family's needs. Maybe your dog likes to sniff through the window, but it's an awkward angle for particular cars

- Making an impulse decision on a car that's too expensive
- Overpaying for the car
- Missing obvious defects, like mechanical problems
- Manipulated paperwork/titles, odometers, or other intentional misrepresentations

You may be thinking of many other risks to consider. You clearly can't predict every single risk event. It's also not always practical to effectively address every single risk you have identified. Your preferred risk management option might adequately address many of the potential problems you identified.

You place a high value on the ability to transport people and pets in this vehicle safely. If any of those risk events happen, you must continue safely transporting those people and pets. Similar to the resilience concept of HROs, here resilience refers to the ability to sustain or restore basic functionality following a stressor. As defined by Erik Hollnagel, a resilient system has the ability to:

- Respond to threats in an adaptive manner
- Monitor what is going on
- Anticipate risk events and opportunities
- Learn from experience

While the definition of resilience is not typically debated, resilience management can mean different things to different applications. Some examples of resilience include: Resilience for engineered systems (e.g., manufacturing facilities and power generation) can relate to initiatives to promote robustness and redundancy. Resilience for organizations can relate to methods to be proactive, train for risk scenarios, and promote situational awareness. Resilience for individuals can refer to the ability to recover from challenging risk events and use the experience to further grow.

Again, it's essential to keep a tight ship. Here, that entails having a constant awareness of what is going on. Once you choose a vehicle, keep it properly maintained and inspected. React immediately if you notice anything unusual, like an unfamiliar sound or reading on the dashboard. Anticipate risk events by doing your homework during the car buying process, park the car in a safe area, treat it nicely while driving the vehicle, and do anything else that's considered a best practice for car ownership. Finally, this car might not be the last car you will ever purchase. Learn from the experience of buying and driving this car and use it to inform future purchases.

As part of building organizational resilience, there is also a need to consider how to respond to various crises. Some best practices include:

- Take responsibility for crises that occur in your domain area. Recovery is much more difficult if problems are hidden and left unaddressed.
- Have a public relations plan. News and social media can be quick to develop narratives about crises and risk events. It is difficult to take control of a story, especially after it has gained reader attention.

- Consider apologizing. The concept of apologizing after a risk event is controversial. Apologies may be avoided at all costs or be underused, they may be required (e.g., mandates, public relations policies), or possibly overused (e.g., practice of "easier to ask forgiveness than permission"). Some may say an apology opens up doors to legal liability. Others say a well-meaning apology can help deescalate a situation.
- Stay prepared for the next crisis. Knowing which parties are responsible for particular crisis-response functions (e.g., communications, mobilizing response, and coordinating among entities) avoids wasted time and energy.

Finally, there is also a need to have humility in recognizing that not all events can be foreseen. Similarly, even with a dedication to resilience and crisis management, individuals and organizations may still not be prepared for every possible event. Agility is key here, as agility allows for quick adaptation to new situations. Well-managed organizations also tend to have clear organizational hierarchies, invest in adequate training for all employees, and maintain the situational awareness necessary to identify risk events when they emerge.

Key Takeaways

- The four risk management options are:

    - Accept the risk
    - Transfer the risk
    - Mitigate the risk
    - Avoid the risk

- Resilience is the ability to sustain or restore basic functionality following a stressor.

5 Risk is not objective

Learning Objectives

After reading this chapter, you will develop further knowledge on aspects of:

- How to consider the integrity of evidence when understanding and managing risk
- Concepts of information, misinformation, and disinformation as they relate to risk

---

One thing that surprises many students of risk is the recognition that we are all risk analysts and we are all risk managers. Every single one of us is responsible for ensuring that risk is managed for our organizations and our lives.

We often share risk-related information with others without consciously thinking about the task. For example, we could:

- Remind children to be careful when running around
- Warn a relative about the risks related to smoking
- Ask a colleague to double-check information shared in a report

We are also often consumers of risk-related information. For example, we could:

- Check the weather forecast in preparation for a hurricane
- Consult safety ratings while shopping for a new vehicle
- Get expert advice before investing money

The question is: What makes risk-related information credible? And how should we protect ourselves from being misled? That is the focus of this chapter.

DOI: 10.4324/9781003329220-7

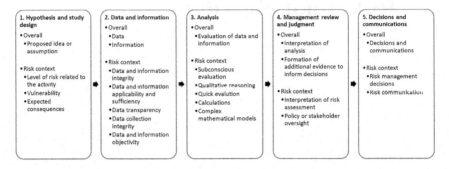

Figure 5.1 Generic process for using evidence as a basis for judgment

Suppose you take a leisurely stroll through your neighborhood on a beautiful sunny day. You see dark gray clouds approaching and sense some concern about taking a walk. You decide to check your phone and find a weather forecast. After seeing the radar showing a big storm approaching, you decide to go home.

Let's break down this encounter into several components, as we describe in Figure 5.1:

Hypothesis and study design: The gray clouds prompt you to investigate whether it is safe to go on a walk. Because of the impending storm, you decide to gain information from various weather forecasts and use those forecasts to make a risk-based or risk-informed decision about whether to go on a walk.

Data and information: You gain information from several phone apps. Presumably, the information came from reputable news agencies. Those news agencies use or purchase data/information collected from sensors, which is then incorporated into complex weather prediction models that are maintained by others.

Analysis: You look at the radar imagery and rain probabilities across those several apps. You perform your own analysis of the situation, determining a high probability of a storm, with strong knowledge support.

Management review and judgment: You interpret your analysis of the weather forecasts to form an opinion: You really do think it's going to storm soon.

Decisions and communications: Because you believe a storm is approaching, you decide to go home. When you see a friendly neighbor outside, you also suggest that the neighbor head indoors too.

Within each step of this process, there are many opportunities to question credibility. Let's talk through each step separately.

Hypothesis and study design: The study design is the roadmap for all of the steps to follow. There are several key components of the study design that ensure credibility:

- Asking a study question: It's important to have a clear question to be answered. The questions could be: "Is it safe to go for a walk?", "Should I start a new bakery?", "Should I buy flood insurance?", or "Which shoes should I wear to the beach?"
- Expertise: Once you have a risk-related question, you should also ensure that you, or the person trusted to answer the question, have the expertise to carry out the process of answering the question. For example, you have vast experience with interpreting weather forecasts, so you feel comfortable carrying out your study.
- Biases: There is a need to minimize biases in your study. Maybe you or someone involved has a pre-determined outcome. For example, while you enjoy taking a walk, maybe you really want to be home to catch up on risk documentaries on TV. We will talk more about biases in the next chapter.
- Reproducibility and transparency: If the approach selected for all components of this study is transparent, you could be able to write down a set of instructions for someone else to correctly follow. If your approach is reproducible, that person would be able to read your instructions and carry out your approach exactly as you would.

Data and information: Data and information are never going to be perfect. There will always be limitations. Here are some essential questions to ask when determining if the data and information used to make your study credible:

- Relevance to the study: The data and information being used should be relevant to the question you're trying to answer. For example, if you're concerned about the weather over the next 30–45 minutes, you consult a weather forecast. Suppose instead, you consulted the yearly average temperatures for your city using a reputable website. Using yearly averages to predict the current weather would be a poor choice for relevance.
- Accuracy: There will always be some reason to question accuracy. Maybe your data is not as current as you would prefer, you have a limited number of sensors, or you're relying on a limited number of observations. Maybe information, such as predictions, is prone to error. For example, while weather forecasts can be inaccurate (e.g., you remember many times the weather forecast was wrong), you feel comfortable using the historical data and the weather forecasts.
- Measurement: Maybe your data or information doesn't actually measure what you're really trying to measure. For example, suppose you accidentally consult the weather forecast from a different city. This issue is particularly troublesome when trying to measure things that are not typically quantities. Suppose instead, you asked your neighbor "How nice is the weather today, on a scale from 1 to 10?" That's a really vague and subjective question. Surveys are prone to this issue.

The format of the survey, the question wording, the ordering of questions, the incentives for taking the survey, and a wide variety of other factors can impact how the question is answered.

Analysis: Analysis can consist of a subconscious evaluation, qualitative reasoning, quick calculations, or complex mathematical models. These analytical methods offer simplifications of reality. Those simplifications can still be very useful for forming opinions and making decisions. There are several characteristics of the analysis that promote credibility:

- Approaches/methods: The approaches/methods should be appropriate for the study. Suppose, instead of consulting multiple weather forecasts, you used a coin flip to conduct your study.
- Implementation: The approaches/methods should be correctly performed. Suppose you intend to compare the rain probability forecast from multiple weather forecasts. Instead, you inadvertently collect and compare the windspeed projections.

Management review and judgment: Reviewing the analysis and generating some opinions is a very subjective task. There are several issues to consider:

- Interpretation: The outputs of the analysis should be interpreted appropriately. While you're able to collect information from multiple forecasts, you may not interpret the information correctly. For example, if app A says there is 50% of rain and app B says there is 75% chance of rain, what is the chance of rain? And what does a "chance" of rain really mean? This is where the rain example breaks down into a credibility issue. Probabilities can be difficult to interpret and communicate.
- Links among steps of the study: There should be a link among all of the steps of your study. The data/information should feed into the analysis, and that analysis should feed into the opinion made. For example, you used the weather forecast analysis to come up with an opinion about whether it will storm soon. Suppose, instead you ignored the information from the weather forecasts. That would signal a breakdown in the links among the steps of the study.
- Context: Everything you need to know about a problem cannot necessarily be understood using the data and analysis. There are a variety of other factors that need to be considered. These factors could involve the needs and concerns of the stakeholders involved, laws/regulations, reputation, and ethical issues. For example, suppose you relied solely on your analysis of the weather forecasts to form a decision about whether to take a walk. You could be neglecting other information, like the length of the proposed walk, whether you should wear a raincoat, whether there are shelters available throughout your path, the time of day, and how any walking companions feel about the walk.

- Values of decision-makers: All decision-makers have implicit or explicit values that must be considered. Values could include financial implications, ethical concerns, issues of health, and safety. For example, suppose your values placed a large emphasis on safety. That emphasis could motivate you to immediately go indoors if your analysis suggested a storm was approaching. Suppose, instead, your values placed a large focus on the health benefits of exercise. You may instead choose to take a walk anyway, but perhaps with a raincoat and a plan to call for assistance if there is potential for lightning.
- Expertise of decision-makers: Decision-makers are not necessarily experts in the subject matter or in risk science. However, they are accustomed to dealing with uncertainty and risk in a practical setting. You may need to consider the need for risk-related explainability to determine how much technical information the decision-makers need to know in order to help them form a stance that can eventually lead to a decision.

Decisions and communications: Once you conduct your analysis and form an opinion, you are now responsible for making a decision and possibly communicating the risk information to others. There are several credibility-related issues to consider:

- Decision: The decision-making approach should have been predefined during the hypothesis and study design phase. The actual decision is based on the management review and judgment.
- Communication: The way information is presented greatly impacts how it is perceived and the effectiveness of the risk message. Consider two scenarios: (1) You casually mentioned to your neighbor that there is an impending storm. (2) You flailed your arms and yelled to your neighbor, instructing them to go inside immediately. In scenario #1, your neighbor might not take the message seriously. In scenario #2, your neighbor may interpret your behavior as irrational, but still, take it seriously and go indoors.

We not only have to be concerned about the integrity of risk-related information we share with others, but we also have to be vigilant about the information we consume. We can classify the information we encounter into three categories: *information*, *misinformation*, and *disinformation*. While it would be nice to base all of our opinions about the world on true and objective information, we often unknowingly encounter *misinformation* and *disinformation*. Even before the advent of social media and the 24-hour news cycle, the concepts of misinformation and disinformation have plagued science, journalism, business, and societies.

Let's first assume that *information* is true and objective. It's free from opinions and judgments. For example, I can look at the weather station in my backyard and determine the temperature and humidity. There will always be some sense of error. I decided to place the single sensor under the apple tree, so the sensor sits in the shade at times, and is also protected somewhat from the elements. The sensor has its own margin of error. If the temperature reading says it's 71 degrees F, and if the margin of error is one-degree F, then I can assume the true temperature is somewhere between 70 and 72 degrees F. While I made a lot of decisions in getting this reading, like the time of day, placement of the sensor, the type of sensor, and so on, there is no intent to deceive by using this temperature reading. Suppose I'm fully aware of the margin of error and all of the other assumptions (placement, time of day, etc.), and I'm comfortable with the inaccuracies introduced by those issues. In that case, I can interpret this weather reading as being *information*.

Misinformation is wrong or misleading while recognizing that falsehood can also be a subjective judgment. *Misinformation* is considered an innocent falsehood, like spreading false rumors in which there is no intent to deceive. Suppose the sensor for my weather station falls in the grass and gives me false readings. The weather station says it's 59 degrees F. If the sensor was hanging in the correct spot, it would have said it was 71 degrees F. Then, suppose I tell my neighbors it's only 59 degrees F. I'm not trying to deceive when I spread that false temperature reading.

Disinformation is wrong or misleading information in which there is intent to deceive. The key here is the word "intent." The intent could exist for many reasons. Maybe the disinformation-sharer enjoys spreading disinformation, or maybe this sharer strategically shares false information for some gain. Suppose the sensor for the weather station fell because a visiting relative knocked it into the grass. Perhaps that relative preferred to stay inside and build model trains instead of spending time in the yard. Fifty-nine degrees Fahrenheit would have been slightly cold for comfortably playing outdoors, while 71 degrees F would have been perfect.

Because risk issues involve judgment, sometimes risk issues are interpreted as *misinformation* and *disinformation*. For example, there are many ways in which *misinformation* could be unknowingly generated. These could include:

- Using new software and making a mistake in calculations
- Taking shortcuts
- Not following instructions
- Outsourcing tasks to someone else who may not be trained or doesn't fully understand the task at hand

It's also possible for risk issues to be *disinformation*. Political agendas, employer meddling, and biases are often reasons for *disinformation* to emerge. Some entities are incentivized to spread disinformation. Examples of *disinformation* could include:

- False advertising to sell products
- False information to influence elections

- Fake scientific research to gain attention for research findings
- False rumors to hurt someone's reputation
- News from media outlets funded by those with the incentive to spread disinformation (e.g., political or social ties)

Consider methods that can be used to support sources of disinformation. Consider the case of a proposed wind farm to be used for electricity generation. In reality, the risk issues are nuanced and complex, with many competing stakeholders and considerations. There is not necessarily a "right" or "wrong" perspective to take on this topic, but instead the perspective should include consideration of the many factors and tools discussed in this book. Disinformation-sharers could begin with the intent to either block or support this initiative using several means:

- Oversimplifying the risk issues, such as by focusing on a single dimension of the risk (e.g., the aesthetics, impact on wildlife, reliability, demand, cost, and community sentiment)
- Seeking quotes from experts with a pre-defined stance or with limited credentials, or paying sources for information (known as checkbook journalism).
- Using quotes that answer leading questions. For example, the writer could ask an expert "Do you think a wind farm is a waste of resources?" While the expert could have an opinion on this, the term "waste" is value-laden and the context does not objectively explain how "waste" is defined. The expert is then asked to provide a definitive answer to the question, which can be used to frame the story. Conversely, the writer could ask an expert "Are you outraged by our lack of investment in wind farms", which is also subject to bias.
- Using non-content-based cues to support the stance. For example, these could include the tone, wording, emotional imagery, font, selection of sources, selection of supporting versus non-supporting details, and so forth. A disinformation sharer could say "LOCAL RESIDENTS SLAM ENERGY COMPANY EXECUTIVES FOR WASTED RESOURCES ON WIND FARM", or alternatively, "Experts OUTRAGED by energy company refusal to invest in clean energy."
- Using exaggeration or sensationalism with simplified or omitted details, creating controversy or division (known as yellow journalism).
- Instilling fear or ether emotions (e.g., rooting for the "underdog," and defense against the "bully").

What's troubling is the gray area between *information*, *misinformation*, and *disinformation*. A risk study that's properly conducted could be interpreted as *misinformation* and *disinformation*. For example, consider the different frameworks for risk management in general. Enterprise Risk Management, ISO 31000, and many other internal corporate policies are well-accepted. However, they are all slightly different. Using any of these frameworks or a slightly different framework from someone else could open up accusations of *misinformation*. As another example, suppose a risk study uses mathematical models that are not appropriate or not credible. Or, the risk study intentionally only considers some dimensions of the risk and omits others. Or, the risk study models use data from a biased source. These issues with the risk study and models could open up accusations of *disinformation*.

We should always stay current by using tools and technologies that help with risk-related activities. There may also be legal aspects to consider, so it's essential to think about legal protection. Employers may not always be incentivized to fully protect its employees. Anyone sharing risk-related information should document their work carefully, be mindful about potential for misinformation/disinformation, and be transparent about all assumptions. If you're ever uncomfortable with how risk is being studied and treated, you should consider speaking up. Being a whistleblower can be critical when the safety and security of people are involved.

Key Takeaways

- It's important to consider the integrity of risk-related information we share with others, and also consume.
- Information is classified using three categories: *information*, *misinformation*, and *disinformation*.
- *Information* is true and objective.
- *Misinformation* is wrong or misleading, while recognizing that falsehood can also be a subjective judgment.
- *Disinformation* is wrong or misleading information which is intended to deceive.

6 We are all biased

Learning Objectives

After reading this chapter, you will develop further knowledge on aspects of:

- How biases impact how we understand, perceive, and react to risk
- How other factors influence our understanding of risk

Suppose you work for a logistics company that ships items across the world. Let's call them Total Business Management (TBM). TBM is known for being a premium shipping provider that employs the use of container transport, air freight, trucking, warehousing, and everything in between. This company has been hit hard by recent risk events such as:

- A ransomware attack that halted all of their operations for a week
- A hurricane that led to delayed shipments
- A workforce shortage impacting their ability to employ drivers for their truck fleet

TBM is not only working through its current problems, but they also plan to be more proactive about dealing with the next risk event. Your team decides to conduct a risk assessment but is concerned about incorporating biases. Similar to any professional risk assessor, the eventual goal is to avoid or counteract the biases discussed further.

The first task is to consider potential cognitive biases. Cognitive biases have been widely studied in both the risk, psychology, and decision-making literature. Many of us encounter our own cognitive biases every day. Let's discuss some biases that commonly apply to risk:

Availability bias: Our memories are not perfect. We may more easily remember some events versus others. We may more easily remember

DOI: 10.4324/9781003329220-8

more frequent events or more recent events. For example, TBM may more easily remember the ransomware attack because it occurred very recently. Or, TBM may more easily remember storm disruptions because they happen so frequently. In a risk setting, in which TBM is attempting to characterize risk and determine what the highest priority risk events are, they may overly focus on those risk events that are more easily remembered.

Anchoring bias: We are often given anchors or initial pieces of information about a particular topic. Anchoring is a commonly used negotiation tactic. For example, suppose you were shopping for a new pair of jeans. The first pair you saw was $300. You become anchored to the price of jeans being $300. Suppose you keep shopping and find a similar pair of jeans for $150. That cheaper pair of jeans would then appear to be a bargain (even if it's not by any other measure). In a risk setting, TBM may be influenced by some initially produced set of risk events to study, sample likelihoods for risk events, or initially provided consequences for these risk events.

Hindsight bias: Once we are aware of the outcome of some event, we may overestimate our ability to have foreseen the outcome. For example, suppose you are adamant about the outcome of a political election. You have a favorite candidate, your friends all support the same candidate, and the polls agree that this particular candidate will win. However, the other candidate wins. In hindsight, you may realize that the other candidate did indeed have a fair chance of winning. In a risk setting, TBM may be surprised by risk events, such as the ransomware attack. They may have previously thought that type of attack was unlikely. However, after the attack, they may realize that they were indeed vulnerable to that type of attack.

Confirmation bias: When we encounter some situation or topic, we often have some prior beliefs or values. We may search for or favor information that confirms those prior-held beliefs or values. For example, suppose you hold firm political beliefs. You find a news source that tends to align with your beliefs. By limiting your news access to that single source, you are selecting to read new information that confirms your prior-held beliefs. Confirmation bias may not only be a manual task. Social media algorithms, search engines, and marketing organizations may also leverage confirmation bias to show you content that aligns with any behavior or personality profiles they have created based on your past behavior and preferences. Your web search results are contingent on how you ask a question, signifying the phenomenon of telling you what you already want to hear. For example, pick your favorite web search tool and type in the following searches:

- What are the benefits of seat belts?
- What are the downsides of seat belts?

Also, try experimenting with the choice of words you use in these searches. You may notice that your search results tend to favor a particular stance on seat belts, depending on the tone and wording of your search.

In a risk setting, TBM may already hold beliefs about risk-related topics. For example, a highly vocal team member may strongly believe in concerns over natural hazard events, like hurricanes and snowstorms. That team member may seek out data and information that can lead to these events being perceived as a relatively higher priority than other types of events.

Self-serving bias: When positive events happen, we tend to take credit for them or believe that our actions were responsible for them. When negative events happen, we tend to blame those events on external factors. In a risk setting, TBM may attribute the ransomware attack to a shortcoming of their outsourced cybersecurity provider. As a result, they may overlook their own role in managing similar risks in the future.

Inattentional blindness: When we focus on certain features, like events or objectives, we may fail to notice other unexpected yet fully visible features. This is a prevalent issue in organizations. It is common to hear that professionals are "busy putting out fires" as they become busy with daily responsibilities. TBM may delay or neglect risk analysis and risk management activities in a risk setting because they are busy with other tasks. Additionally, TBM team members may focus on risks they encounter in their daily professional responsibilities and neglect those occurring, or only noticed, in other departments or job functions.

Mirror imaging: We tend to assume that our adversaries think like us. In reality, the adversary may have different values, varying risk tolerances, may not be rational, and may want to deceive us. Various organizations and governments have different bureaucracies and leadership styles. In a risk setting, we may recognize the challenges of predicting the actions and behaviors of others. TBM may be confident in its abilities to incentivize workers to join their team, alleviating a labor shortage, but their predictions could be wrong.

In addition to the individual cognitive biases, some biases develop in team dynamics. These can be equally harmful in a risk setting. A few of the more widely known biases are:

Group-think: Sometimes, groups of highly qualified individuals form a consensus on some viewpoint or conclusion, but that consensus is not rational or held by all group members. Group-think may lead to a stifling of controversial issues, critical evaluation, and creativity. In a risk setting, group-think can be very dangerous. Team members could be peer pressured into forming some conclusions that overlook important issues with considerable risk implications. For example, vocal members of the TBM team may view the risk process as a perfunctory exercise and neglect to discuss prominent issues like black swans and resilience-building. Other members of the team may go along with the substandard risk activities due to group-thinking.

Authority bias: In group settings, there is often a mix of seniority or power among team members. That power is assumed to be supported by organizational hierarchy, but could sometimes be influenced by social factors (e.g., tone, body language, and fear). Team members may place a relatively larger weight on the opinions of those authority figures. Those authority figures may be more experienced and have stronger knowledge, but this is not always the case. In a risk setting, there is not necessarily a guarantee that the authority figures are more knowledgeable or more correct about particular facets of the studied risk issues. For example, an executive in a large oil and gas corporation may have a strong understanding of financial performance but may have little understanding of day-to-day operations, such as those related to compliance with safety policies. Team members with on-ground experience with safety policies may question safety policy compliance assumptions and hesitate to speak up about noncompliance. This bias can severely impact the team's ability to characterize risk. TBM may limit their risk discussions to topics that have been pre-determined as being a high priority by executives on the team.

Courtesy bias: Team members may limit their opinions to more socially acceptable ones, thereby avoiding offense or controversy. This bias may be more prevalent in particular fields, organizations, or cultures. In a risk setting, team members may avoid sharing potential risk factors in order to avoid bringing up unpleasant topics. TBM may avoid studying risks that relate to their personal sense of health and safety because they are uncomfortable discussing them. They may also avoid discussing particular types of risk in effort to avoid offending the individuals and teams responsible for addressing those risks.

In addition to the biases impacting team dynamics, general issues with teamwork may impede risk efforts. Problems like poor leadership, insufficient conflict resolution, conformity, ineffective communication, and many other types of poor group dynamics are commonly encountered. Insufficiencies in team dynamics can cascade throughout risk analysis, risk management, and risk communication activities, eventually compromising stakeholders' safety and well-being.

The overarching question is how to avoid the biases and deficiencies in team dynamics. There is no simple answer to this question. A trained risk analyst would be aware of these biases and attempt to minimize the impact. Here are some strategies that are typically encountered in the risk profession:

Outsource risk activities: Sometimes, organizations do not have sufficient training and resources to carry out risk-related tasks. They can bring in neutral third-party consultants to help develop risk standards and complete risk-related activities.

Develop a risk culture: A risk culture consists of shared beliefs, norms, values, practices, and structures, with respect to risk, in an organization. A strong risk culture implies that a risk team takes accountability for completing all necessary steps for risk analysis, risk management, and risk communication practices. If team members notice biases or poor team dynamics emerging, there should be safety in speaking up and addressing those issues.

Peer review: In academia, there is a concept of peer review. A new research idea or analysis needs to be evaluated and vetted by a neutral third party. In a risk setting, organizations may seek to have their risk activities reviewed various organizational stakeholders or choose to have an external third party conduct a review. This review can identify missed ideas and validate the beneficial aspects of those risk activities.

Be aware of general workplace dynamics: Are there any known issues with managing the organization? Or how do stakeholders impact one another? Deficiencies in general management principles can leak into other facts of the organization, such as quality, risk, reputation, and employee sentiment. A solid risk culture relies heavily on a well-managed organization with healthy workplace dynamics.

Quick Exercise

Let's do a quick exercise. Imagine a scenario where you are planning a beach vacation in paradise. You have eagerly planned the details. You have plane tickets, hotel reservations, and friends ready for the big trip.

You go to bed early the night before the trip. You are sure to get a full night's rest so you have the energy to spend the day traveling. Then, at 1 am, you awaken and realize that you didn't adequately plan for risk. What are you worried about?

• Take a minute and make a list of the risks that concern you the most.
• What did you write?

Here are some items mentioned by students who attempt to answer the same question:

• Missing the flight
• The effects of intense exposure to ultraviolet rays at the beach
• Sharks
• Bacteria and viruses
• Navigating new roadways on the way to the local attractions
• Forgetting to pack something important (e.g., toothbrush and clothes)

Your list might not match the one shown earlier. You may have additional items or you might not be worried about some of the items listed here. That's OK. We all have different perceptions of risk.

For each of the risks you wrote down, determine a rating on a scale from 1 to 10, denoting your level of concern. A rating of 10 implies the highest concern while a rating of 1 implies the lowest concern.

Here are some common answers seen among students:

- Missing the flight: 10
- The effects of intense exposure to ultraviolet rays at the beach: 2
- Sharks: 9
- Bacteria and viruses: 2
- Navigating new roadways on the way to the local attractions: 5
- Forgetting to pack something important (e.g., toothbrush and clothes): 2

Next, try to explain your reasoning for how you rated each risk.

Using the ratings as shown earlier, we can ask why some risks were more concerning than others. The concern about missing the flight might be large because it's 1 am on the night before the flight. The concern may not have been rated as high a few weeks before the flight because, back then, there was more time to prepare. Bacteria and viruses may be rated relatively lower because we constantly deal with this risk and it's not specific to this beach vacation. The issue of sharks might be very concerning because shark attacks are particularly scary.

Some of these ratings don't align well with the likelihood of these particular events occurring, as reported by general statistics. For example, suppose the death rate from a shark attack is around 1 in 3.5 million during one's lifetime. Also, suppose the likelihood of dying from the flu is about 1 in 60 during one's lifetime. Yet, the ratings shown earlier imply the concern about shark attacks is drastically larger than that of bacteria and viruses. While we might have statistics about which of these risks is most prevalent, our perceived risk doesn't always match our statistical risk. Our perception of risk is often influenced by factors such as personal opinions, past experiences, societal expectations, knowledge, and timing. Also, these statistics (e.g., probability of dying of a shark attack) refer to the general population. Your likelihood may be drastically different. For example, the likelihood for a surfer who frequently spends time in shark-infested waters might be very different from someone who doesn't go anywhere near those areas.

There is abundant scientific literature that studies how we process information related to risk. The following list describes some widely studied factors that influence how people perceive and respond to risk:

- *Dread versus No Dread:* We tend to judge a risk as being higher if it elicits emotional responses like fear and terror (e.g., terrorism, shark attacks); and we tend to judge a risk as being lower if it does not elicit an emotional response (e.g., recreational activities like mountain climbing).
- *Involuntary versus Voluntary:* We tend to judge risk as being higher if it involves activities that are involuntary (e.g., contaminants in foods); and we judge risk as being lower if it involves activities that are voluntary (e.g., smoking).

- *Uncontrollable versus Controllable:* We tend to judge risk as being higher if it involves activities that we ourselves don't control (e.g., flying in a plane controlled by a pilot); and we judge risk as being lower if it involves activities that we control (e.g., driving ourselves in a car).
- *Unfamiliar versus Familiar:* We tend to judge risk as being greater for activities that are unfamiliar to us (e.g., new hobbies); and we judge risk as being lower and more acceptable for activities that are familiar (e.g., driving in a car).
- *Human failure versus Nature:* We tend to judge a risk as being greater if it is generated by some human failure or incompetence (e.g., industrial accidents); and we tend to judge risk as being lower if it is caused by nature (e.g., solar radiation).
- *Personal versus Impersonal.* We tend to judge risk as being greater if it impacts people personally (e.g., an earthquake fault line in our own backyard); and we tend to judge risk as being lower if it does not impact people personally (an earthquake fault line in a sparsely populated and far away region).
- *Non-equitable versus Equitable*: We tend to judge risk as being greater if it is viewed as being unfair (e.g., population inequities related to a pandemic); and we tend to judge risk as being lower if it is viewed as being fair (e.g., seat belt use mandated for all passengers)
- *Irreversible versus Reversible*: We tend to judge risk as being greater if the consequence is irreversible (e.g., loss of endangered species); and we tend to judge the risk as being lower if it is reversible (e.g., injuries with an expected short recovery).

The aforementioned list shows that there are underlying factors that can help us further understand how we perceive risk. We can look to the literature describing those underlying factors. Daniel Kahneman popularized the terms *System 1 versus System 2.* System 1, deemed the experiential system, is an automated, quick, instinctive, and emotional thinking process. System 2, the analytic system, is a slower, logical, and more deliberate thinking process.

In a risk setting, we may have some quick involuntary rules or judgments about interpreting or reacting to risk scenarios, much of which is controlled by System 1. We instinctively know that touching a hot stove or hot pan will cause pain, so we choose to be extra careful around the stove. Maybe we have experienced picking up hot pans in the past, so those experiences have shaped how we handle that particular risk. Once System 1 has developed, it can be a conscious effort to change those risk-related behaviors.

In a risk setting, our thought process to handle new and complex risks is largely handled using some combination of System 1 and System 2. System 2 can build logical connections between some new risk and other more simplified risks. System 2 may require more evidence to form some conclusion, and that evidence needs to be supplemented by some form of analysis. System 2 thinking is largely connected to the risk analysis methods employed by risk managers following risk science principles. Professional risk assessments are based on System 2.

The need for System 2 in professional risk assessments, combined with the biases and issues with cognitive processing presented in this chapter, reminds us that risk is not all about statistics and objectivity. Risk is about judgments. Risk perceptions could lead to resources being used in "non-objective" ways (e.g., using disproportionate resources for a risk that would only have a marginal impact), but we can also learn from these perceptions. These perceptions do represent real concerns of individuals and organizations. These perceptions may represent some vital aspect of risk that's not fully captured in a risk assessment.

For example, consider risk perception biases between involuntary versus voluntary activities. Suppose a community group is highly opposed to the risk associated with a new cell phone tower. This is not an unusual opposition, as the term "not in my backyard" or NIMBY commonly refers to opposition to development near one's home, while the opposition may not exist if placed further away. In this case, residents claim the cell phone tower will emit radio frequency waves that could have negative health effects. This community group considers this risk involuntary, as they are not self-selecting their location to be a site for this new cell phone tower. Despite no strong evidence that exposure to these radio frequency waves causes serious health effects, communities remain concerned. The community risk perceptions are high, while experts do not recognize this risk. There are uncertainties and one should be careful concluding that these perceptions are "just feelings" that do not reflect important aspects of risk. This remains a valid concern that should be addressed using stakeholder dialogue. There may be sources of misinformation regarding the science of radio frequency waves, other non-health reasons for this opposition that need to be addressed, and other measures that could ease community concerns.

Because of the biases and factors involved with our perceptions of risk, knowledge of risk science is essential for properly understanding and managing risk. Also, acknowledging the cognitive factors influencing our perception of risk, we recognize that communication of risk messages is incredibly challenging. The next chapter will take more detailed look at the topic of communication.

Key Takeaways

- A wide variety of biases impact how we perceive and react to risk.
- Due to these biases, our own characterizations of a risk may not agree with data-based statistical characterizations.
- Despite issues with biases, these perceptions reflect the real concerns of individuals
- Perceptions of risk, with or without biases, can include important aspects of risk not fully captured by risk assessments.
- System 1 refers to the automated, quick, instinctive, and emotional thinking process.
- System 2 refers to the slower, logical, and more deliberate thinking process.

7 The message matters

Learning Objectives

After reading this chapter, you will develop further knowledge on aspects of:

- Purposes for risk communication
- Characteristics of effective risk communication
- Characteristics of misleading risk communication

Risk communication consists of the sharing of risk-related messages (data and information). For example, communication efforts may involve a state health agency sharing critical safety-related instructions to the public. Communication may involve a corporation discussing risk issues with shareholders and employees. Communication may also involve information exchange among a variety of stakeholders, such as individuals, community groups, private industry, regulators, and the media.

Risk aspects that are often communicated include public health, such as managing risks related to contagious diseases. Communication messages may also relate to consumer safety, such as related to informing the public about misuse or side effects of products like prescription medications. Communication can also be critical during emergency management activities. For example, communication before,[1] during, and after a hurricane can be vital for purposes of preparedness, evacuation, and recovery efforts.

The general goals of risk communication in a public context consist of informing communities and seeking community input on potential risks. Through this process, community members are able to understand the risk further and form their own risk-related opinions. Typically, or ideally, risk communication efforts result in reduced overall risk. These suggested activities are often behavioral (e.g., mask-wearing, not drinking contaminated water, etc.). In the long run, consistent risk communication can promote trust

DOI: 10.4324/9781003329220-9

and credibility, allowing the public to address risk more effectively, promote community safety, and facilitate recovery after a risk event.

You may not be fully aware of the many risk communication messages typically encountered every day. For example, consider messages like:

• Social media memes telling you how you can protect yourself during a pandemic
• Warning labels on products (e.g., cigarettes, alcohol, cleaning products, fast-food coffee cups)
• Warning labels on shared objects (e.g., gasoline pumps and grocery store carts)
• Traffic signs/warnings (e.g., marking of hurricane evacuation routes)
• Safety data sheets and signage in industrial settings
• Public meetings to discuss new infrastructure investments

Other communication methods are not as common these days but could include telephone calls, mailings, and community workshops.

We can figuratively compare the warnings in risk communication initiatives to roadway traffic circles (also known as roundabouts). While drivers either love or hate these traffic circles, the design of the traffic circle intends to require drivers to slow down and more consciously make driving decisions. Similarly, risk communication messages attempt to catch the attention of our System 2 (discussed in the last chapter) as we make risk-related decisions. Some of these warnings and messages seem unnecessary, such as the warnings on the fast-food coffee cups. Fear of lawsuits can contribute to this abundance of warning messages. Too many warning messages might also lower our sensitivity to the most important and critical safety-related messages.

The broad goal of risk communication is to help stakeholders obtain an increased understanding of risk. Then, individuals can form their own opinions. However, if organizations rely on shortcuts or misguided suggestions, individuals develop their views on limited and sometimes untrue information. Communication is not as simple as sharing data or instructing people about how to think or behave. People perceive risk differently, and people perceive risk messages differently.

Any type of communication message contains two components. First is the topic of sensation, which involves our sensory receptors (eyes, ears, nose, etc.). Then, there is assessment and perception, the process by which the audience interprets those sensations. A communication message can give those sensations meaning.

When we encounter messages, we go through the process of information processing. We only process information we notice. So many neurological and cognitive factors determine what information we process and how we process that information. This relates heavily to the biases we discussed in the previous chapter.

In the risk field, message saliency is essential. Saliency can involve getting someone's attention and also keeping someone's attention. Many human factors relate to the saliency of a message, and many of those factors can work on concert with each other to improve that saliency. Here are some ways message saliency relates to sensation and perception:

Vision: Message elements, such as fonts, colors, and format, can influence how we view a message. A method to attract attention can involve choice of colors, contrast, or use of unusual or unexpected visual elements. Colors can have emotional reactions, as studied in the field of color psychology. While widely contested, some believe that colors can be used for therapeutic purposes, impact test performance, and can help market products to customers. Some examples of theorized emotional reactions to colors include:

- Red: Danger, strength, and energy
- Green: Nature, harmony, and freshness
- Blue: Peace, authenticity, and trust
- Pink: Gentleness, optimism, and sincerity
- Purple: Power and ambition
- Brown: Reliability, ruggedness, and simplicity
- Black: Security and formality
- White: Safety and cleanliness

Colors can potentially impact risk perception. For example, red may be perceived as the riskiest, yellow is the mid-range, and green is the least risky. Many have formed theories about how colors impact risk and safety. For example, it has been asked whether blue streetlights could reduce crimes, whether pink walls could reduce aggression in prisons, or whether bright colors in hospitals could increase anxiety and disrupt patient recovery.

The design, including the font of a message, can also make a message visually salient. For example, using capital letters or exclamation points can attract attention. Symbols are also widely used to quickly and effectively share safety-related information.

Smell: Odors can stimulate emotions and memories. Odors can be used for customer marketing for air fresheners and foods. Odors can also be used for purposes of risk and safety. For example, a chemical that smells like rotten eggs is often added to natural gas to make natural gas leaks recognizable, allowing for a timely evacuation and avoidance of potential natural gas disasters.

Sound: Sounds can also influence emotions and behaviors. For example, many studies have attempted to find associations between genres of music and the emotional or physical response of listeners. In a risk setting, sounds can be critical for conveying danger messages. For example, fire alarms and tornado sirens provide a high-volume warning of emergencies.

Touch: Tactile stimulation can influence how messages and physical objects are perceived. Touch is widely used in marketing contexts. For example, tactile stimulation is used in product packaging, body care products, services, and any products intended to be touched. In a risk setting, we see touch being used to make safety devices accessible. Accessible pedestrian signals provide nonvisual cues to signal whether it is safe to cross the street. These nonvisual cues include audible tones and vibrotactile surfaces, which are combined with visual cues.

Taste: Our taste receptors can also influence how we experience a product or situation. Of course, taste is widely leveraged in the food industry. In a risk setting, taste can be used to avoid potential poisonings. For example, a bittering agent is often applied to antifreeze and coin batteries to avoid the disastrous health effects of consuming those products.

There is emerging guidance on risk communication. The research community, government agencies, and the private sector are developing creative and innovative solutions. The United States Environmental Protection Agency (USEPA) offers the following seven cardinal rules for risk communication:

Rule 1: Accept and involve the public as a legitimate partner
Rule 2: Plan carefully and evaluate your efforts
Rule 3: Listen to the public's specific concerns
Rule 4: Be honest, frank, and open
Rule 5: Coordinate and collaborate with other credible sources
Rule 6: Meet the needs of the media
Rule 7: Speak clearly and with compassion

These seven cardinal rules suggest that risk communication is much more than a single message. These rules show the value in creating partnerships with the community. This partnership is not meant to replace action but to develop solutions to risk-related problems. Also, effective risk communication does not assume a homogenous audience and messaging strategy. Different subsets of the public have different needs and concerns. Honesty, trust, clarity, and compassion are vital to effective risk communication.

Risk communication is essential in emergency settings. The communication can consist of emergency preparedness activities, messages during an emergency, and also messages that support recovery. These communication messages enable populations to make informed risk-related decisions. The effectiveness of these communications depends on the level of trust between the communicator and the communities.

Those risk communication messages are not to be seen in isolation but integrated with other emergency preparedness and response activities. Communication is often combined with community activities. For example, strategies of talking about risk and safety early in primary school education can help children understand risk and encourage parents to actively participate in risk-reduction activities.

There is a constant need for continuous review of the effectiveness of risk communication. Also, no emergency is the same.

Communities are likely to have many questions, concerns, and fears. Giving the community resources to address those fears before misinformation and rumors influence them can be a critical first step.

Increasingly, social media is highly leveraged for risk communication efforts. Social media is often used for engaging with the public, situational awareness, gauging public concerns during an emergency, and addressing rumors of misinformation. However, those communication efforts should ideally be used consistently even before an emergency in an effort to gain trust and credibility.

Language is also important for emergency communication. Some typical features of effective risk messaging include:

- Clearly and simply state the response (e.g., move to an interior room during a tornado threat)
- Coordinate with different information sources to provide consistent, non-conflicting information (e.g., emergency management agencies coordinating with news agencies, sending a consistent message)
- Use simple language (e.g., avoid highly technical terms or complex statements)
- Consider social and cultural contexts (e.g., ensure the message is accessible by all audiences)
- Use credible information to support the message (e.g., scientific papers, news, and government agencies)

In times of emergencies and other chaotic settings, implementing these features of risk messaging requires coordination among the message sharers. For example, consider a local news station sharing critical weather-related information during a tornado emergency, while that coverage is constantly interrupted by less useful warning messages sent by the local cable provider. While the local news station and cable provider have intent to effectively communicate, the conflict among the messages contributes to overall risk, as individuals cannot access the information they need.

As mentioned earlier, using credible information to support a risk message is essential. However, information sources that are credible to one person may not seem credible to another person. The issue of information credibility alludes to the generally problematic factor in risk-related communication, which is misinformation, misrepresentation, or intent to deceive.

There are many reasons why fake messaging can emerge in a risk setting. For example, scammers may take advantage of emergency situations in an effort to trick their victims into making payments. Journalistic or political motives can give a platform for false risk-related statements to be used for political gains or increased readership of articles. Efforts to address fake information have been largely ineffective in the past. People always have to be vigilant about fake messaging, but this is increasingly important in a risk-related context.

These concepts and issues with risk communication apply to all settings, not only public sector risk. Risk communication at the enterprise level is also important in ordinary operations and atypical or emergency situations. Consider some typical scenarios in which risk communication can be used in an enterprise setting:

- Meetings with shareholders. These meetings could include discussions about risk events and scenarios that are of high concern. For example, a quarterly earnings call could discuss the potential for an upcoming recession.
- Social media messages (postings or user groups). These messages (text, video, infographics, memes, photos, etc.) could have some two-way dialogue, as the general public can weigh in on current risk-related issues. For example, a social media post could provide opportunity for the public to comment about issues with quality control, shipping, service, impact on the environment, and societal impact.
- Blog posts. Blog posts are often a form of a press release in which customers, investors, and the general public can self-select into accessing a corporate message. For example, a company could create a blog post to discuss its newly advanced sustainability initiatives. Usually, this is one-way communication. However, this can be shared on social media in an effort to spread the message and gain feedback.
- Advertisements. Advertisements are also a form of a press release. Print advertisements could be in newspapers/magazines, direct mail, billboards, and so on. These types of print advertisements are most often one-way communications. TV and online advertisements could use email, website, social media, or video formats. The technology-based advertisements allow customers to be targeted using collected preference-based and demographic information. These can be one or two-way communications. For example, a company could purchase a page in a local newspaper to address rumors about the company or its operations, which is one-way communication. An example of two-way communication through advertisements would be posting video advertisements on websites that allow the public to comment, and responding to those public comments.
- Press releases. Press releases can be used by journalists to develop news stories. For example, a press release could be used by a pharmaceutical company to present the latest scientific breakthroughs. Journalists can make their own interpretations of the content or tone of the press release.

Newer technologies and communication methods can sometimes make risk communication efforts challenging to gauge. Entities (companies, societies, individuals, etc.) have the ability to purchase their audiences and even purchase feedback to their risk communication messages. Some mechanisms include purchasing "followers" on social media sites. In addition, companies can send free products to customers, who then can comment on social media or write reviews on e-commerce websites. Companies can also pay for endorsements from internet celebrities, journalists, and social media influencers.

In these cases, followers can be incentivized to provide positive feedback in two-way communication attempts, which can mask risk issues. Reviewers can also be incentivized to remove negative comments with risk implications. Communication platforms may remove, reduce visibility, or reduce the weight of comments with risk implications. Some examples of comments with risk implications could include customer feedback about food safety problems, community concern over environmental impact, complaints about product safety, or employee complaints about worker safety. Whether those types of comments are true, untrue, a systemic risk problem, or something in between, they still can provide useful information about risk, risk perception, and stakeholder concerns.

These types of arrangements can complicate the ability of the public to interpret those risk communication efforts. Some case examples include:

- A company demonstrates their environment-friendly practices and uses paid endorsements to collect mostly positive responses from users claiming to be educated on issues, encouraging others to take the same stance. For example, if an oil and gas company seeks to share their new greener and renewable energy initiatives, they can use this strategy to gain a more positive image.
- A corporation takes a stance on a public health concern by using their social media following. In the social media posts, the corporation discusses risk-issues related to the public health concern. For example, a company taking a strong stance on containing the COVID-19 pandemic, seeking fewer adverse health outcomes, could post on social media messages to encourage mask-wearing and vaccinations.
- A corporation or nonprofit lobbies for regulation. The public may see messages about the lobbying efforts and also contact their elected leaders. Examples of risk-related lobbying actions could include lobbying for mandatory sustainability reporting or lobbying for stronger worker safety regulations.
- A company encourages their employees to follow worker safety rules.
- A company attempts to gain approval for a new facility that may have some environmental, health, or community-related impacts.

Outside of typical operations, atypical situations in risk communication may be necessary for an enterprise setting. Consider the following risk-related scenarios:

- Your company just endured a cyber-attack, losing the personal information of millions of customers. Your company needs customers to take protective action immediately.

- Your company's manufacturing facility is accused of contaminating their site. Executives did not know about this issue. As a result, your company faces a significant public relations crisis combined with substantial efforts and regulation-related activities to address this contamination.
- While progressing smoothly to date, your construction project faces a potential supply shortage that could cause the project to become far behind schedule and encounter high unexpected costs. You have to explain this issue to the project funders.

The goal is not to wait for risk events to happen but to think about appropriate policies and responsible parties for risk communication. Companies and individuals can adopt some of the principles described earlier in this chapter, but also develop their own internal policies that follow each organization's spirit, values, and mission.

In any atypical risk situation, the public, employees, and other individuals look to those in charge to answer difficult risk-related questions and to have a strategy to address the risk. In situations where there is a high degree of uncertainty and the highest-paid or highest-ranking person in the room has no idea what to do, this can create a sense of tension and distrust. In these situations, wrong or harmful guidance based on little evidence or knowledge can still appear misleadingly credible to individual audiences. These situations can also lead to a loss of control, as others can attempt to take advantage of these situations for personal gain.

Risk communication efforts, whether in the public sector, corporate, or another type of setting, should also be grounded on high-quality risk science. If the risk is understood poorly even by the leaders of the risk study, the communication efforts will also likely be poor.

Let's talk through some more specific examples of false messaging. Suppose you saw the billboard in Figure 7.1.

Figure 7.1 Example of false messaging

Of course, we can all approach this billboard with an ounce or two of skepticism. One may ask questions like:

- Who created this billboard?
- How was this number determined?
- How does the creator define "fake news"?
- Why is the creator sharing this information with me? What is to be gained?

These are all essential questions that we will dissect. Even before concepts of fake news and misinformation made news headlines, there has been intense discussion about the integrity of information.

Back in 1954, Darrell Huff wrote the classic book *How to Lie with Statistics*. This book mentions many mechanisms that lead to intentional or accidental non-truths. Many of those mechanisms serve as red flags for identifying misinformation. Many modern sources exist for identifying misinformation. A few of the most commonly encountered mechanisms include:

Issues with the source: We should always consider the message creator. The credibility of a source often helps us evaluate the message's credibility. In the billboard example, there is no source cited at all, which makes us seriously question the presented information. By questioning credibility, we can ask questions like:

- Is a specific source mentioned in the message? A reputable source would likely identify itself as a source for a message. There may be other signs, like copyrights and publication dates.
- Is the mentioned source the actual source? False identities are increasingly easy to create. For example, maybe you get phishing messages in your email in which some email-sender pretends to be someone you know in order to get information from you. Maybe you have been a victim of someone stealing your photo and identity on social media.
- Was the message creator incentivized to share this information? Maybe the message creator is trying to get the readers to act in some way. Perhaps the actions could be buying a product, voting for a particular political candidate, being aware of some health issue, or behaving in some way. Not all incentives are bad, but some malicious actors may make it difficult to know if you were being used or manipulated in some form.

Issues with the message: Often, the way messages are presented is just as important as the message itself. Issues to consider include:

- Was the message value-laden? Sometimes manipulative messages try to get readers to react emotionally. They can get readers riled up using a variety of mechanisms, like tone, font, body language, poignant imagery, and emotionally-charged language.

- Is the message as simple as the topic? We often consume information in bite-sized pieces. However, many of the big issues in the news are not so simple. These topics have multiple perspectives, rules/laws, and contexts. A message that tries hard to simplify these complex issues may be trying to sway you to take a particular stance.
- What does everyone else say? Are multiple news sources saying the same thing? Are the news sources that agree with one another on the same ends of the political spectrum? Was the article peer-reviewed if the information is in an academic journal?
- Was the supporting information disclosed? Reputable sources would explain the sources of data and information. They should also include other explanatory information, such as assumptions and data collection methods.

Issues with graphics: There are all sorts of tricks that can lie or mislead with data. For example, consider Figure 7.2 showing a hypothetical organization's estimated credibility (scale from 0 to 100) for three different news sources. This is a poor figure for several reasons. First and foremost, the axes are not labeled. Also, notice the drastically lower credibility for Source B. According to this figure, Source A appears to be more than twice as credible as Source B, just judging by the heights of the bars.

Notice something wrong with the figure?

Notice that the y-axis begins at a rating of 82. Starting at a value that's not zero messes with our sense of scale for the bars.

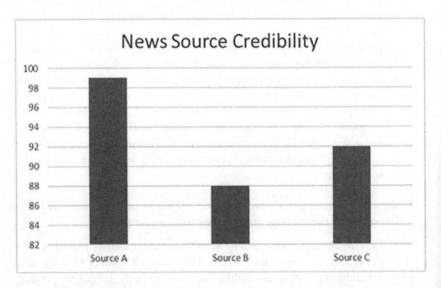

Figure 7.2 Credibility for three news sources

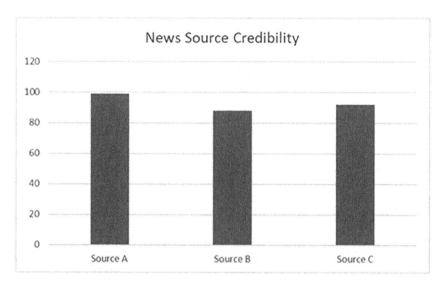

Figure 7.3 A corrected figure showing credibility for three news sources

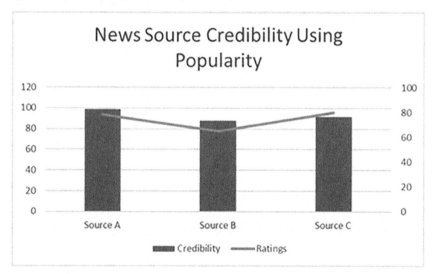

Figure 7.4 Credibility for news sources and their popularity

Figure 7.3 shows a partially corrected figure. Now, the *y*-axis begins at zero. Now, judging only by the heights of the bars, Source B doesn't look quite as bad as it did before.

Figure 7.4 further complicates the message. Now there is also a secondary axis. The *y*-axis on the left corresponds to the credibility rating, with each of the bars showing that credibility rating. The *y*-axis on the right corresponds to the estimate of popularity for the news source (scale from 1 to 100), with the line height showing that popularity metric. While you may be able to infer some relationship, there is too much wrong with this figure to make a proper assessment. Let's think through all the issues:

- The axes are unlabeled, making the figure confusing.
- The axes don't show units, and there is no real explanation for where the numbers came from or what they represent.
- The scales were from 1 to 100, while the y-axis shows 120 as the maximum.
- The continuous line implies that there is some space to fill between each news source. The news source data is categorical (containing only three items and nothing else), so showing the line information between news sources is misleading.

The lesson of this example is that we should always remain vigilant about figures that can mislead. Pay attention to issues like axes, colors, font sizes, and spacing. These aspects can make an image pretty on a screen, but they can also distract you from objectively viewing the information.

The assumption of random sampling: Whenever we encounter statistics, we state some point using data. For example, we can study 100 news articles, decide which ones are fake news, find that 67 of those 100 are fake news, and turn that into a calculation:

$$\frac{67\ fake\ news\ articles}{100\ news\ articles\ studied} = 67\%\ of\ news\ aricles\ are\ fake\ news$$

We would call those 100 studied news articles the "sample" in a statistics course. We're building that 67% statistics using only that sample data. If we had the time and energy to study all of the news articles ever published, or a very large set of news articles, that dataset would be called the "population." A perfect analysis would have sample data that is representative of the population. In real life, a representative sample can be challenging to achieve. We often refer to a perfect representative sample as a "random" sample, such that each member of the population, or every set of members in a population, is equally likely to be selected for the sample. Any features suggesting a lack of a random sample opens up the possibility of including bias in the calculation.

Suppose the creator of this billboard collected data. Their entire dataset of 100 news articles was sourced from a single news source. This particular news source was known for taking a comedic stance on recent news topics and often intended to stretch the truth. This was definitely not a random sample. If this were a random sample, the creator could have selected an equal number of articles from each news source while having to decide about which news sources to include. Those articles would have been randomly selected from each of the news source websites. Note that the billboard did not mention any data collection or sampling methods.

> *Sample size matters:* Just as it's important to have a representative sample, it's equally important to have a sufficiently large sample size. The trouble is that data collection can consume a lot of time, energy, and financial resources. For example, if you're conducting a survey, you need to find a sufficient number of people to fill out the survey, which takes a lot of effort. You may need to incentivize the survey-takers with some sort of reward, like cash, which can be costly. On the billboard, there is no mention of sample size.

In statistics courses, students are often reminded of the magic number of, meaning that a sample size of 30 opens up doors to statistical testing. We thank the Central Limit Theorem for helping us understand that the accuracy of statistical estimates increases with larger sample sizes, which opens up many statistical opportunities. In the age of big data, sufficiently large sample size can be easier to find in many contexts. For example, with public data on social media, we can easily collect enormous datasets.

> *The method matters*: Having a random sample and a sufficiently large sample is also not enough. We make many assumptions when we use analytical methods, including statistics. We don't always check those assumptions or communicate limitations associated with assumptions and data. Additionally, even the most vetted statistical methods can be misused (e.g., p-hacking), making it even more difficult to identify misused methods.

Even with a perfect message, there are many challenges with effective risk communication. These challenges include:

> *Ambiguity*: Scientific studies may not always agree with one another. In fact, studies are deemed publishable when they offer some type of contribution or new finding. When there is disagreement in science, those disagreements can be used to counter or question science-based risk messaging.
> *Politics*: Many risk topics have become incredibly politicized. Consider issues like global warming and mitigation measures for the COVID-19 pandemic. The political discourse around risk topics based on science can undermine the effectiveness of risk messages. For example, risk messages for politicized topics can lead the angriest of citizens to dig their heels in more to do the exact opposite of the risk message.
> *Habits and strongly held beliefs*: Policymakers, government agencies, and companies have tried to influence behavior by using messaging. Examples include printing calories on menu items to address the obesity epidemic

and printing warning labels on cigarettes. The effectiveness of these messages is questioned mainly because these messages are intended to counteract habits and strongly held beliefs.

Message sharing: A risk message is not effective if it does not reach the intended audience. Social media, marketing, and search engine algorithms make it increasingly difficult for audiences to encounter viewpoints that differ from their currently held perspectives.

All of these challenges are not in isolation. We may find that these challenges interact with one another and make it increasingly difficult to develop effective risk messages.

As a result of these factors, risk communication is not only about singular messages that are shared with the public. Basic communication can involve conveying data and information to the general public in an attempt to transfer knowledge. Communication efforts can also attempt to persuade other stakeholders (e.g., the public) and try to influence behavior (e.g., smoking, alcohol, seat belt use) or address stakeholder concerns (e.g., food safety, nuclear installations). It's important to note that risk communication can involve a two-way communication process in which the public is invited to inform risk decisions and vice versa. But the reality is that risk can also involve debate, which means risk communication can lead to increased disagreement and distrust.

In simplified situations, communication involves information transfer, such as sharing the results of a technical study, evidence, and so forth. The communication can include accessing an audience and sharing information.

In cases where there are issues with the trustworthiness of a risk-related issue, there may be a need for dialogue with stakeholders and the public.

Risk amplification and public distrust can emerge in cases of immense disagreement over risk topics. Roger Kasperson and colleagues developed the social amplification of risk framework (SARF), which describes two levels of risk amplification: The first happens in processing risk-related information, and the second occurs in societal responses. When the risk is communicated in a message, that message interacts with other processes to either intensify or weaken the perception of risk and its manageability. Risk amplification happens when there is an intensifying or increase in importance or volume of risk signals, leading to a heightened perception of risk and consequently triggering risk reduction activities. Conversely, risk attenuation refers to a weakening of the importance or volume of risk signals, leading to a lowered perception of risk and a reduction of activities to reduce risk.

An example of risk amplification involves large press coverage of a risk event, such as a predicted natural disaster or terrorist attack. This coverage can lead to a greater emphasis on reducing the impact of these events, calling for individuals to mitigate risk. This amplification could also result in social disorder, changes in regulation, or promotion of public distrust about this risk.

An example of risk attenuation involves scientific research demonstrating a lower than previously thought level of risk. Examples could be new research demonstrating lower-than-thought levels of cancer risk resulting from alcohol consumption or reports of declining crime rates, which could result in reduced regulation and a reduced level of public distrust. Risk aspects could then be ignored.

Attenuated risks may appear to be sufficiently managed, appearing in the public eye to be a reduced risk that can be attributed to proper leadership and management. However, in reality, an attenuated risk can result in a lowering of risk reduction activities, resulting in insufficiently managed risk. In the long run, risk attenuation can lead individuals and organizations to not respond or under-respond to attenuated risks. Consequently, those attenuated risks could become amplified risks in the future.

As an example of fluctuating risk amplification and attenuation, consider risk related to color additives in foods. These color additives have a long history of being used for the production of processed foods, drugs, and cosmetics. They also have a long history of being banned for various safety-related reasons. These additives can be useful from a risk and safety perspective (e.g., visual aids for correct identification and dosage of medicines), and they can be useful for marketing products (e.g., visual cues for candy/soda flavors, making food appear more appetizing). One particular additive, red dye 40, has received a lot of media and political attention over the last couple decades. This dye, used in candies, soft drinks, and medicines, has been studied as a possible contributor to hyperactivity in children. Significant media attention and global regulatory interest in banning the additive contributed to risk amplification. This amplified risk led to greater public and regulatory interest in understanding the implications for food additives in general. However, additional high-profile research questioning the link between the additive and hyperactivity, and reversal of product bans, led to attenuation of the risk. With lowered public attention toward the risk, this attention frequently becomes amplified following new research findings, fluctuating media attention, and increased availability of natural-dyed food products.

These issues of risk amplification and attenuation remind us of our limited capacity to concurrently think about the many risks we encounter. The very act of prioritizing risk for mitigation initiatives is a conscious effort to focus attention on some risks instead of others. The palette of risks we prioritize could frequently change as particular risks are amplified and attenuated, even when we recognize that our risk behaviors are influenced by factors of risk perception. Risk messaging efforts can also adapt to these factors of risk amplification and attenuation. For example, these factors could help the message sharers choose the appropriate tone, urgency, and frequency of messages as they are directed toward particular populations.

If there are more significant issues with public distrust and an imbalance of values and world views, there may be a need for dialogue and mediation. Examples of dialogue and mediation could include public hearings, surveys, focus groups, advisory committees, conferences, panels, and mediation. These various methods can explore the audience's concerns, allow the audience to engage with one another, and lead to dialogue that can promote trust and engaged problem-solving.

Key Takeaways

- The general goals of risk communication consist of informing communities and seeking community input about potential risks.
- Risk communication is also equally important for nonpublic sector applications, including corporations, nonprofits, and societies, and individuals.
- High-quality risk communication requires high-quality risk science.
- If risk is poorly described and understood, risk is also communicated poorly.
- Methods of risk communication can include stakeholder dialogue, warnings, signs, and social media messages.
- Issues with risk communication and information reliability can be impacted by issues with misinformation, misrepresentation, and intent to deceive.

Note

1 Now would be a good opportunity to take inventory of your own safety and risk initiatives. Resources could include health or emergency management agency websites. Note their main themes of communication. Many agencies will also provide suggestions for preparedness, like having disaster preparedness plan and supplies for various types of risk events. The agencies may also have suggestions for activities to promote overall safety (e.g., smoke detectors, carbon monoxide detectors, and water safety).

8 Risk is not just about dollars

Learning Objectives

After reading this chapter, you will develop further knowledge on aspects of:

- History of financial performance-based metrics in management
- How additional nonfinancial metrics are increasingly being incorporated into overall strategies
- Issues and fairness and ethics as they relate to both financial and nonfinancial metrics when applied in a risk setting

This book has emphasized that risk science involves things we value. Many of us value all kinds of things, including health, safety, financial resources, our property, and many other aspects. One aspect that often gets the largest weight is life.

Thinking about life using monetary units can seem disturbing and uncomfortable. However, "Value of a Statistical Life" (VSL) is a very commonly used term in healthcare, insurance, and economics. While VSL is often misinterpreted as putting a dollar value on human life, it instead serves as a tool for understanding and managing risk to improve health and safety outcomes.

VSL can be used for understanding the benefits of avoiding fatalities, with the eventual goal of being used for decision-making and comparison of risk-reduction policies. In the past, the value of a statistical life was based on a person's expected earnings. Some VSL estimates consider other factors, related to characteristics that can be used to describe individuals and populations.

For example, a VSL estimate in a transportation agency is used to understand the monetary value of preventing fatalities. In other words, VSL is the cost that individuals would be willing to pay for safety improvements and reduced risk that would decrease the expected number of fatalities by one. This VSL estimate is not the value of a single life, but the monetary value applied to the reduction of risk.

DOI: 10.4324/9781003329220-10

The VSL value varies from application to application. However, typical values in US dollars range between $5 million and $20 million. The estimate may be impacted by inflation, income growth, income elasticity, and other factors. The VSL within a particular organization or profession may change yearly.

In addition to VSL, there is also consideration of nonfatal injuries. Instead of death, these types of injuries result in some loss in quality of life. This loss can present itself in reduced income and also pain and suffering.

To account for quality of life, another calculation that is commonly used is the Quality-Adjusted Life Years (QUALY), representing the measure of life expectancy and quality of life. It is often used to measure the cost-effectiveness of new treatments for improved health outcomes. Insurance companies can use it to decide on appropriate medical treatments and pharmaceutical products coverage. These metrics can complement the Maximum Abbreviated Injury Scale (MAIS) on a scale from 1 to 6 (Minor to Un-survivable). Using this information, evaluate risk related to particular types of injury.

What is helpful about metrics like VSL and QUALY is that they can be used for cost-benefit analyses. Companies and agencies use these estimates to understand values for risk reduction. These analyses can be used to compare willingness to pay for risk reduction from new treatments, policies, and other risk mitigation measures.

As you can imagine, the idea of VSL and cost-benefit analysis is not without controversy. However, they remain prevalent because they provide a common metric that can be used across applications (environmental agencies, healthcare, insurance). They can also be explained quantitatively with set guidelines and measurements.

Because the phrasing of VSL is often misinterpreted as putting a dollar value on a human life, other alternative names have been explored. These include "value of prevented fatality," "value of improved chance of survival," "accident reduction cost," and "willingness to pay to reduce the risk of dying." These terms have more positive undertones, and better represent the intent of computing the metric: to understand how to best address and treat risk.

VSL is convenient for demonstrating efficacy in policy-related decisions and regulatory initiatives, but is also highly controversial. VSL is criticized for focusing on a single element of risk: mortality risk. As we have discussed in this book, there are also other things we value that contribute to the overall quality of life. Those other aspects we value are not as easily quantified, making them more challenging from a risk aspect. VSL is also criticized for over-simplifying risk in general. VSL in a cost-benefit analysis can overly focus on single policies that can be measured by considering short-term versus long-term implications.

To address the shortcomings of VSL, many alternative focus areas have gained popularity in recent decades. While these focus areas are not as mathematically convenient as VSL when evaluating costs and benefits for risk, they significantly improve risk-related discussions and decisions.

The term "Triple Bottom Line" has become popular in recent years. In the past, businesses and government agencies relied on financial costs and benefits when evaluating performance and making risk-related decisions. Those financial metrics made it difficult to fully consider implications for things like environmental health and well-being of people. The three components of the triple bottom line triplet are:

- People: People includes all stakeholders (not only shareholders for corporations). These stakeholders can include employees, customers, suppliers, and community members. The impact on these people can vary by application, but can include issues with as poverty, diversity, health, safety, and human rights.
- Planet: Concerns about global warming, pollution, sustainability, and other planet-related factors.
- Prosperity: Economic opportunities and growth are vital for risk and safety. Efforts to address people and the planet require resources. Prosperity includes financial performance for businesses, jobs with living wages, and economic growth.

The United Nations, in their 2030 Agenda for Sustainable Development, developed the United Nations Sustainable Development Goals. All countries are tasked with addressing these goals in ways that can improve global quality of life. Businesses and academic institutions are also building these goals into their overall strategies. The meeting of these goals has the potential for treating major world risk issues. The UN Sustainable Development goals are:

- No Poverty
- Zero Hunger
- Good Health and Well-Being
- Quality Education
- Gender Equality
- Clean Water and Sanitation
- Affordable and Clean Energy
- Decent Work and Economic Growth
- Industry, Innovation and Infrastructure
- Reduced Inequalities
- Sustainable Cities and Communities
- Responsible Consumption and Production
- Climate Action
- Life Below Water
- Life on Land
- Peace, Justice, and Strong Institutions
- Partnership for the Goals

While the goals provide 17 different areas, they are also closely related. For example, the issue of ending poverty cannot be seen isolated from goals to address inequalities, education, hunger, and health, as these aspects influence one another.

The UN describes several "levers" that can be used to address these goals. These levers include governance, individual and collective action, science and technology, and economy and finance. What is notable is the requirement for all of these levers to operate in partnership. We as individuals have a major role in this partnership, as our collective actions can promote change by:

- Understanding the managing risk associated with these factors in own lives and communities
- Supporting leaders who promote regulatory adherence to the goals
- Directing our investments to companies that align with our environmental and social values

A related framework from the United Nations of the Sendai Framework for Disaster Risk Reduction describes actions to reduce disaster risk. This framework consists of seven goals to be achieved by 2030:

- Reduce global disaster mortality
- Reduce the number of affected people globally
- Reduce direct economic loss in relation to GDP
- Reduce disaster damage to critical infrastructure and disruption of basic services
- Increase the number of countries with national and local disaster risk reduction strategies
- Substantially enhance international cooperation to developing countries
- Increase the availability of and access to multi-hazard early warning systems

One of the main challenges with achieving these goals is partnership. There are many challenges in forming productive partnerships among the various stakeholders. For example, partnerships among government agencies can be challenging when integrating information systems, communicating across different workplace cultures and bureaucratic systems, or promoting employee collaboration. It's also challenging to encourage partnerships between industry and government, as they can have very different objectives and operating procedures.

Private industry is also increasingly gravitating away from using singular financial metrics in decision-making. These non-financial metrics could relate to their own internal corporate values, political factors, or be based on what are the current norms in their industry. These non-financial metrics have potential to spur innovation, create brand strength, attract customers, attract highly qualified and productive employees, attract investors, and reduce public criticism. However, there is also possibility for unintended consequences and controversy (e.g., financial infeasibility, inadequate implementation of initiatives, boycotts, and reputational issues).

One example of a movement to understand the corporate impact is using the B Corporation Certification. An assessment of a B Corporation includes dimensions of:

- Governance: Including elements of mission and engagement, and ethics and transparency
- Community: Including elements of diversity, equity, and inclusion; civic engagement and giving; supply chain management; and economic impact
- Environment: Including elements of environmental management, water, air and climate, land and life
- Customers: Including the element of customer stewardship

With the momentum toward positive impact, there can be efforts to "greenwash." Greenwashing occurs when some entity shares misleading information about their efforts toward addressing the risk-related issue of sustainability. While greenwashing efforts involve values and communication, two of the core facets of risk, these efforts can also be political and controversial. Risk science generally does not have a view on the political aspects of corporate values and practices like greenwashing. However, many examples of greenwashing exist. For example, companies have been accused of greenwashing through the use of language in promotional materials, despite historical issues with environmental impact and contributions to global warming. Even using the color green in logos and promotional materials can make a company appear more environmentally sustainable. Food companies have been accused of greenwashing in attempt to make processed and pre-packaged foods appear healthy and sustainable. Fast fashion companies have been accused of greenwashing to make their products appear as sustainable options, when the items equally contribute to overall clothing waste and offer only a marginal improvement to sustainability in production.

With the generation of any new product, service, or project, there is also a need to consider performance metrics across the entire life cycle of the product, often called a life cycle cost or whole-life cost. The components, measured across any performance metric of choice, not limited to financial metrics, represent the total cost of some asset (product, service, etc.) from purchase through disposal. While the components of this cost can vary by type of product and the use of the product, cost components often include some combination of:

- Purchase/Financing
- Design
- Building/Installation
- Operation
- Maintenance
- Depreciation
- Disposal

Customers often focus on the purchase cost, neglecting the other components. As a result, there can be unintended consequences in which customers are unaware of the overall risk concerns associated with the product.

A classic example of this is the captive product pricing model. This model has been used to sell items like coffee makers, razors, and home printers. Customers who focus on purchase price may purchase a product with a low upfront pricing, but neglect to see costs associated with ownership and use. Consider the case of a coffee maker sold for a relatively low price. To produce coffee using the coffee maker, customers must purchase additional products. This type of pattern can be harmful for nonfinancial metrics also. Consider the environmental impact of the containers used for individual coffee servings.

On a larger scale, consider decisions related to the purchase of a new electric vehicle. There are many factors to consider, such as the upfront costs of purchasing an electric car versus the costs of replacing batteries for those cars, whether the electricity generation and distribution needed to power the electric car is "cleaner" than if a more traditional gas-powered motor was used, the environmental impact of the materials being used for the batteries, or the carbon footprint of a newly produced vehicle versus that of already-produced vehicle on the road (and not disposed/scrapped too early). It could be asked how the environmental benefit of the new vehicle compares to other environment-friendly lifestyle changes over time (e.g., food choices, food waste, driving habits, and home insulation). The authors of this book do not intend to voice a stance on the answers to those questions, but instead seek to demonstrate that there are many factors to consider and they are not easily simplified into yes versus no or good versus bad stances. All of these factors have risk-related implications.

As a whole, this discussion of this chapter brings to light issues with ethics and justice in a risk context. The heart of risk revolves around the opportunity and the avoidance of harm. In general, harm is a worsening aspect of something we value.

However, analyzing and managing risk requires investment of resources, often money and time. This investment is in pursuit of avoidance of harm or pursuit of opportunity. For example, consider the post-9/11 policies of airline safety and security. We all invest our time and energy into security checks and increased surveillance for the sake of overall global safety.

In a risk setting, the measurement of potential harm is represented by the metrics we choose, as discussed in this chapter. However, addressing those metrics we choose can induce further or additional harm across other aspects. For example, risk policies and procedures may have unintended consequences that can cause additional harm (e.g., decreasing safety and negative impact on the environment). Consider the hypothetical example of a football team wearing advanced helmets that provide increased protection against head injuries, including concussions. These helmets are important and vital for protecting players from serious and debilitating consequences from head injuries. However, the helmets can also promote a false sense of security for those players.

Consequently, the players can take more physical risks and encounter more serious injuries to other parts of the body.

The football example could also exemplify a moral hazard, in which the football player increases their exposure to risk (e.g., riskier behavior) because they do not take on the total cost of the risk (e.g., buffering by the helmet). Moral hazards can also exist in many other settings, such as when homeowners build homes in hurricane-prone waterfront areas, while buffering that risk with insurance and rebuilding assistance.

Harm can also emerge from developing risk policies to address those metrics we choose, but measuring those metrics only for the overall population, without considering various segments of that population. For example, consider laws to address jaywalking. In many cities, individuals may face tickets and fines for crossing a road in areas other than traffic lights and crosswalks. On the surface, this is a pragmatic approach to promote safety, by ensuring drivers are trained to look for pedestrians in those locations. However, these jaywalking laws also limit areas where pedestrians can cross streets, causing them to walk far away from their shortest path to find a safe crossing place. Those pedestrians, without cars, face greater walking distance, a potentially larger safety risk walking on those roadways, and can face heavy fines. Thus, the risk policy may have helped on the surface when viewing overall metrics, but decreased the safety for populations without vehicles, including some of the most vulnerable individuals.

Harm can also emerge from developing risk policies to address the most important metrics, without recognizing how those new risk policies can impact other metrics, which are also largely important. In other words, this relates to developing risk policies without realizing important tradeoffs and related unintended consequences. For example, consider the case of developing scorecards for patient care. Individual physicians, departments, and healthcare facilities can be graded on their performance. These scorecards could consist of objectives for quality patient care, measurements, targets, and opportunities for comparison among healthcare facilities and physicians. These metrics could include patient satisfaction (through surveys) or more specific information on patient outcomes. For example, scorecards for kidney transplant facilities may share statistics on the number of transplants conducted, the time to receive a donor transplant, and the one-year survival rate for patients.

In some cases, if certain performance targets are not met, and if the information is publicly available, patients may elect not to use a particular facility. In other cases, government agencies may intervene, and the healthcare facility can face serious repercussions, potentially facing closure of the facility. This could result in many unintended consequences, such as:

- There could be an increase in healthcare costs, causing at-risk patients not to be able to afford life-saving procedures, thereby deteriorating the overall health of populations.
- Individual physicians or healthcare facilities refuse to treat high-risk patients, out of concern for maintaining required performance levels.

- Those healthcare facilities that offer life-saving treatments to high-risk patients appear to be performing poorly, but instead are offering critical care to patients who have few alternative opportunities for treatment.
- The most qualified physicians for high-risk patients seek employment in higher-paying facilities, which may not be in locations where those high-risk patients seek medical care
- Burnout or job insecurity for employees could be seen in healthcare facilities with poor performance levels, leading to lower quality healthcare.

Information about various metrics of concern is also becoming more publicly available. One can think back to histories in which risk-based decisions happened in closed environments where only the decision-makers knew the context and the data/information being used to form those decisions. For example, consider the issue of guidance for vaccinations. Historically, patients did not typically have data and information about vaccine-related research. However, with the increased use of the Internet and freedom of information, patients can readily access seemingly unlimited information from sources, including academic journals, web resources, agencies tasked risk communications, businesses, journalists, political groups, influencers, celebrities, scientists, and also the unscientific opinions of others.

These sources may attempt to prescribe risk-related decisions for their audiences. However, as we discuss throughout this book, individuals make their own decisions based on the available knowledge and their own values. These values of individuals can conflict with the values of those offering guidance for overall populations. The future frontier of risk science involves the question of freedom, such that there is a need to consider the balance of trust in individuals to make risk-related decisions for themselves, particularly as individuals have varying values and a varying amount of credible knowledge and training on particular risk issues. Additionally, given this variability, there are ethical concerns when determining the amount of coercion that can be used to promote particular risk management activities.

In relation to concerns over ethics and values, there are additional dimensions to consider. There are many definitions in the social science literature that help to explore the concepts of *equity* or *equality*. We interpret equality as risk policies that seek for each individual or group to be given the exact same resources or opportunities. We interpret equity as risk policies, which recognize that each person has different circumstances, therefore needs different resources to reach an equal outcome. In a risk setting, equity and equality can promote drastically different types of discussions, decision-making methods, and risk management initiatives.

Some examples of equity versus equality in a risk setting include:

- If a public health threat has a disparate impact on various populations, what are the appropriate risk reduction policies and investments?

- If workplace safety outcomes are different for various stakeholder groups, is this a concern? How can this be addressed in ways that meet regulatory requirements and also promote overall safety?
- Is there a way to determine whether the doorways, signage, and design of a workplace are appropriate and safe for all employees? While the workplace may meet requirements from various regulations (e.g., Americans with Disabilities Act and others), is there a need for additional dialogue with employees and outreach with experts to determine the most effective practices?
- Are any social structures in place that tend to harm or help the overall risk in an organization? For example, are there issues with diversity, equity, and inclusivity that need to be addressed? Or, as another example, are all workers invited to voice concerns over risk and safety, without encountering retaliation or penalty?
- Are there any issues with the accommodation resources in an organization, such that the health and safety of employees or stakeholders is compromised? For example, are certain stakeholders subject to particularly severe stressors?
- Are all stakeholder groups included in discussions about risk and safety? Are there groups that are excluded? Why?

The issues of equity and equality, such as those shown in the examples above, are also commonly discussed and debated by political parties. The authors of this book do not intend to declare a particular stance on these topics. Instead, we aim to open up dialogue and encourage readers also to form their own educated stances on these issues.

Finally, it's important to consider issues of justice in a risk context. We can generalize to say there are three types of justice:

- Distributive justice: Fairness in the distribution of something among groups.
- Corrective justice: Fairness of the response to a wrong (e.g., punishments and penalties) to a person or group.
- Procedural justice: Fairness in the gathering of information or how a decision is made. This includes fairness in processes, transparency, the opportunity for voice, and being impartial in decision-making.

The topic of fairness is an important component of values related to risk. Individuals have different views on fairness, but the topic of fairness has been widely studied. Distributive justice in a risk context could emerge as an issue with risk being unfairly distributed across populations. For example, the jaywalking example found an imbalance of risk among pedestrians and those driving vehicles. This type of justice also emerged during the COVID-19 pandemic during which some individuals, such as those working in particular types of professions, encountered higher exposure and consequences during the pandemic. Of course, readers may have their own opinions about the fairness of this situation, as issues of equity can be viewed as a political or economic stance.

Corrective justice also emerges in a risk context, as when things go wrong in a risk setting, there is the question of who is responsible. This has been witnessed as corporations performing acts that cause harm to individuals and the natural environment. The corporations may face fines and other penalties. Individual employees acting on behalf of the corporation may also face penalties and potential legal ramifications.

Procedural justice contains the primary intents of a risk science approach, aiming for effective stakeholder considerations, risk communication, and the reduction of biases. Consider the placement of a new wastewater treatment facility. Likely, communities may be unhappy with the potential for a new sewage plant in their neighborhood. These plants have the potential for foul smells and poor aesthetics. Yet, these facilities are necessary for communities. There may be broad legal requirements or general precedence for public participation in understanding risk and decision-making with these new plants. There can be a questioning of procedural justice if risk and safety-related decisions are made without public involvement. The risk process can promote procedural justice through understanding the community's social values and concerns through open discussion, citizen input, engaging in compromise, and maintaining transparency.

From a broader perspective, the discussion of this chapter suggests that greater attention to risk related to the things we value has many secondary benefits, including greater awareness and discussion of risk-related issues, value generation, and generally improved understanding of risk. In other words, greater attention to nonfinancial risk is critical for things we value. However, this attention toward nonfinancial risk needs to be addressed with a holistic perspective, recognizing the role of ethics and fairness, which can be value-laden and political, in the approaches.

Key Takeaways

- VSL serves as a tool for understanding and managing risk, in pursuit of improving health and safety outcomes.
- The Triple Bottom Line consists of three components: People, Planet, and Prosperity.
- Risk decisions involve issues of ethics and fairness.

Rule 3

Risk can be a good thing

Rule 3 discusses the importance of adopting a risk science approach. While the majority of the risk discussions in this book and in general revolve around understanding negative outcomes, risk science also applies to positive outcomes.

Chapter 9 will discuss the balance between downside risk and also opportunity.

Chapter 10 will discuss individual and organizational challenges in developing a risk mindset. It will also discuss some frequently encountered questions and concerns related to a risk science approach.

Chapter 11 will discuss the empowerment that comes with addressing risk at both an individual and organizational level.

Chapter 12 will discuss case studies as they relate to a risk science approach. The goal is to use these case studies as a model for your own application of risk science.

Chapter 13 will discuss how readers can learn more about risk science.

DOI: 10.4324/9781003329220-11

9 To manage risk is to balance the positive and the negative

Learning Objectives

After reading this chapter, you will develop further knowledge on aspects of:

- How risk management and performance management relate to one another
- The importance of balancing the potential for both positive and negative outcomes

---

Let's briefly remind ourselves of some often-used definitions of risk, as discussed in previous chapters. Many definitions reference some future activity. Risk is then related to the consequences of this activity, with consideration of something that humans value. For example, some common definitions include:

- The possibility of an unfortunate occurrence
- The deviation from a reference value and associated uncertainties
- Uncertainty about and severity of the consequences of an activity with respect to something that humans value

The consequences of the risk activity are in relation to some reference value (some status quo, current conditions, or benchmarked value). Metrics that could be used to measure risk could involve:

- Severity of consequences of some activity and the associated probabilities
- The expected value of consequences
- Consequences and uncertainty characterization associated with those consequences

Judgments of the strength of the knowledge supporting these metrics should always be included.

DOI: 10.4324/9781003329220-12

The key here is recognizing that some of the common definitions of risk focus on undesirable or negative consequences, but others cover both the positive and negative consequences concurrently. Not all risk scientists and fields agree on these definitions, but acknowledging the potential for gains can open up possibilities for concurrently thinking about opportunity and loss.

Those positive consequences are often thought to be *performance*, relating to the functionality of a system (e.g., number of widgets produced, number of customers served, number of defects, productivity, and revenues). Performance management involves business activities conducted to meet some objective related to that functionality of the system (e.g., produce x widgets and achieve a revenue of $1 million). Performance management aims at creating value, but can also seek to avoid accidents related to aspects such as health, safety, and the environment.

While the risk discipline has gained in popularity over the last 40 years, the performance discipline has established roots dating back to the late 1800s and early 1900s. Frederick Taylor, in the early 1900s, was known as a promoter of scientific management. This type of management style had a strong emphasis on efficiency and rewarding workers for productivity. Henry Ford, also in the early 1900s, was known for paving the way for decision-making and investment that promoted efficiencies and productivity. Ford not only developed and refined the assembly line process that simplified the work of assembling automobiles but also leveraged the power of wages to attract productive and skilled workers. Peter Drucker, in the mid-1950s, called for workers to share responsibility for setting objectives, then tracking worker performance using those pre-determined objectives, known as the Management by Objectives regime. Similarly, Quality Management regimes, gaining and growing in popularity from the 1950s onward, call for continuous improvement.

While these management methods and motivation practices can be questionable by today's values and current management practices, what's important is that these performance management practices continue to evolve to address gaps and insufficiencies while also reflecting societal values. More current performance management practices include nonfinancial and non-efficiency-based metrics, like consideration of the many stakeholders and ethical concerns. However, the integration of performance and risk remains a relatively new concept.

Performance has varying definitions across industries and contexts. Some common definitions are:

- The output of a system in reference to what is valued
- Quantified fulfillment of the organization's mission
- A specified system output, uncertainty judgment associated with the output

Different organizations, industries, and contexts have differing measurements for performance. Some examples include:

- Profitability
- Revenue
- Quality
- Efficiency

The definitions of risk and performance show strong parallels with one another. They both involve some future projection influenced by uncertainty and knowledge. With the commonality of definitions of risk and performance, protection and value generation can be balanced. However, in many organizations, risk management and performance management are seen as siloed or separate activities.

A simple argument shows the conflict between risk and performance. For example, there are two perspectives that are commonly taken:

Perspective 1: Strong focus on value generation increases health, safety, and environmental risks. This could be true for organizations with weak management and a weak risk culture. A strong focus on the "bottom line" could encourage an organization to incentivize decision-making and behaviors that pursue short-term gains. As a result, there may be an incentive for taking shortcuts that degrade the ability to address health, safety, and environmental risks. For example, consider a company that introduces a new business activity with huge potential for profit, but the activity involves unaddressed safety risks. As another example, consider an industry involving activities that can be dangerous for workers and the natural environment. Suppose, in pursuit of value generation, workers are offered a bonus based on productivity. As an unintended consequence, workers take shortcuts in their work activities, not following the established safety procedures. In addition, workers are reluctant to report injuries due to the potential for lost time, lost productivity, burdens of paperwork, and any other impacts on their financial bonus.

Perspective 2: Strong focus on health, safety, and the environment can contribute to value generation. While this may not always be true, there are arguments for why this focus on health, safety, and the environment can promote innovation, improved performance, and eventual value generation. Some examples of this type of dynamic include:

- Avoidance of reputational damage that comes from risk incidents
- Improved employee morale through a shared sense of mission that reduces harm
- Improved public image that allows the organization to hire employees who share in the mission

- Improved opportunities for innovation when considering additional health, safety, and environmental constraints
- Avoidance or reduction of risk accidents that could result in severe consequences

Perspective 1 may provide short-term gains, but Perspective 2 can lead to meaningful change and risk reduction in an organization.

Even the youngest of citizens are exposed to the need to carefully balance short-term value generation versus risk. Dr. Seuss' Lorax examines the managerial and ethical quandary of producing thneeds from truffula trees at the expense of the forest, barbaloots, and other forest creatures. We can reframe this to an example of Perspective 1 versus Perspective 2 in the forest industry.

Modern sustainable forestry practices attempt to balance value generation with risk related to biodiversity, local communities, climate change, worker safety, and many other risks associated with logging. While these sustainable forestry practices are met with severe criticism, they are one step in a growing movement to manage risk in this industry. Certification for sustainable forestry can provide several long-term benefits, including:

- Increased monitoring of tree and forest health, generating new knowledge for global environmental understanding, with impacts across industries
- Increased traceability of supply chain activities
- Increased documentation, in cases of accusations of illegal logging
- Community-building for foresters, promoting further innovation and knowledge-sharing
- Sustainability in practices, promoting longer-term employment and health for rural small businesses and communities
- Protecting biodiversity and rare species

Indirectly, certifications for sustainable forestry serve to educate consumers about the risk related to forest products. This can call for customers to think more carefully about the products they consume, recycle, and discard. As the movement toward sustainable forestry grows, it could be followed by an increase in regulatory requirements, customer expectations, and the prices consumers are willing to pay for certified products.

In the case of sustainable forestry. Perspective 2 can potentially provide long-term gains for individuals and businesses. While the act of certification comes at some short-term cost or loss, that loss can be recouped in the longer term.

The pursuit of Perspective 2 incentivizes organizations to view risk and performance as a single organizational process. Table 9.1 shows how performance management's benefits relate directly to risk management's benefits, as commonly implemented in organizations. Generally, performance management is essential for meeting performance-related objectives or targeted performance

Table 9.1 The benefits of performance management and how they relate directly to the benefits of risk management as commonly implemented in organizations

Aspect	Benefits of Performance Management	Benefits of Risk Management
Broad Goals	Decisions that promote high-level performance, to *maximize positive consequences*	Decisions that maintain high-level performance and avoid failures
Basis	Management using evidence and data-based metrics	Use of evidence and data-based metrics, but *with additional inclusion of health, safety, and environment context*
Guidance for Objectives	Setting of performance objectives based on shareholders and direct stakeholder values	Setting of performance objectives to meet direct and *indirect stakeholder values*
Objectives	Well-defined quantitative objectives	Well-defined quantitative and qualitative objectives
Decision-Making and Processes	Decision-making and processes to meet well-defined quantitative objectives for achieving high-level performance	Decision-making and processes to meet well-defined quantitative and qualitative objectives for *avoiding, reducing, or recovering from negative consequences*
Pursuit of Goals	Pursuit of high-level goals	Pursuit of high-level goals, *with a focus on uncertainty in meeting those goals*

levels. In effect, performance management aims to maximize positive consequences. Risk management involves maintaining the meeting of those performance-related objectives, while also considering uncertainties.

Figure 9.1 also illustrates the difference between performance management and risk management. Performance management relates to achieving performance goals. Risk management relates to understanding and managing risk for the impact of potential risk events (losses in performance), recovering, and achieving those performance goals. In the figure, it appears as if there are singular measures of performance. However, real individuals and organizations likely have many competing measurements of performance. As a result, performance management and risk management involve tradeoffs among those various measurements.

The issues of performance and risk even are even further simplified as shown in Figure 9.2. Using this re-visioning of performance and risk, we can define performance-risk as a deviation in performance that is relative to a reference level, such as a performance objective. If future performance, current performance, or historical performance is above that reference level, we can refer to it as upside performance (risk). Conversely, if performance is below the reference value, we call this downside performance (risk).

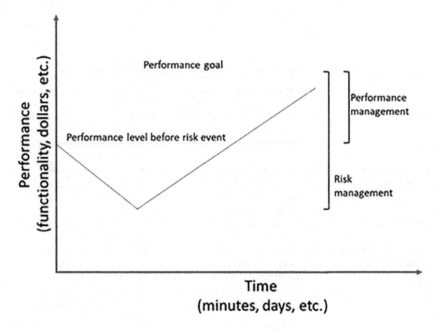

Figure 9.1 Performance management versus risk management

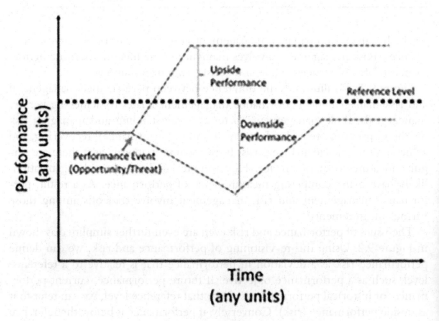

Figure 9.2 Upside and downside performance relative to some reference level

There may also be disagreement on how performance should be measured in general. Shareholders may measure performance using metrics such as revenue, costs, customers, efficiencies, safety-related incidents, and other quantitative metrics. Public infrastructure managers may measure performance using more broad and nonfinancial metrics, like safety, ability to meet demand, and environmental impact.

A branch of applied mathematics, known as decision-support, or multi-criteria decision-making can help organizations work through those complex and potentially competing measures of performance.

Beyond mathematics, several practices can help combine performance management and risk management. Some practices include:

Include stakeholders: Sometimes, managerial decision-making only includes input from those tasked with decision-making responsibilities. Instead, those decisions should ideally involve a wide variety of stakeholders, such as the employees, the general public, communities, and so on. An effective performance-risk program also requires buy-in and commitment from those stakeholders.

Consider low-likelihood events: Considering seemingly unlikely, but high or extreme consequence events is not new to performance-related decision-making. However, this topic is often only taken seriously by the most risk-averse organizations and decision-makers. It can be a beneficial exercise to make it a norm to consider the effects of those seemingly unlikely events within decision-making processes.

Have resilience plans: While we recognize that we can't adequately foresee the future, we can invest in initiatives that promote resilience and agility after a risk event has occurred.

Recognize antifragility: This concept, defined by Nassim Taleb, suggests that shocks, variation, and uncertainty in a system can eventually lead to improvements in the system. Decision-makers can consider how to take advantage of these shocks in ways that introduce system improvements.

Focus on process improvement: No organization is perfect. There are always ways to improve and innovate. A commitment to improve processes, whether pristine or not, relies on accountability for deficiencies in processes and an organization that is willing to change and improve.

Think beyond regulations: Regulations and rules are minimal standards. A truly performance-focused and high-achieving organization can set their own policies and standards while meeting the criteria set by various regulations and rules. The bare minimum should not necessarily be the goal.

Focus on knowledge: As discussed throughout this book, an over-focus on probabilities and consequences for risk events can distract us from understanding the core issues at hand, which is related to uncertainties and knowledge. Low-knowledge versus high-knowledge situations may call for different decision-making conversations.

A commitment to integrating performance management and risk management may open up conversations about protection versus value generation. Protection involves investment. For example, if a transportation agency invests in improved bridges, additional lanes, and emergency management services, those investments come at a high cost. Those investments require not only implementation costs but also full lifecycle costs from design, to implementation, through eventual closure. However, those costs can be seen as a form of insurance against negative consequences if a natural disaster were to impact the transportation network. These investments might have some opportunity costs, as those funds could be used elsewhere for additional value generation.

In the long run, an integrated perspective on performance and risk can be embedded in an overall risk culture. These best practices for developing a balance between performance and risk are also best practices for management in general, resulting in more focused conversations and better-informed decisions.

Key Takeaways

- Common definitions of risk focus on undesirable or negative consequences, but others cover both the positive and negative consequences concurrently.
- The definitions of risk and performance are strongly linked as they both involve some future projection that is influenced by both uncertainty and knowledge.
- Real applications involve many competing measurements of performance, and consequently, performance management and risk management involve tradeoffs among those various measurements.

---

10 Risk management is worth the time and energy

Learning Objectives

After reading this chapter, you will develop further knowledge on aspects of:

- Organizational and interpersonal issues as they relate to developing an effective risk culture
- Questions, concerns, and overall resistance to the creation of risk programs; and how to address this type of feedback

Effective risk programs, policies, and practices rely on a strong risk culture. Many different organizations have different definitions of risk culture. In general, a risk culture consists of shared beliefs, norms, values, practices, and structures concerning risk in an organization. This risk culture guides organizations toward understanding, analyzing, managing, and communicating risk.

We as people are born with some sense of risk culture, and that sense develops over time through experiences, guidance, and rules. While children may not always follow our warnings and rules around risk and safety, children, in general, don't want to get hurt and understand that some activities are riskier than others. Children in a playground often understand that they could fall if they lean too far on the edge of their treehouse.

On the other hand, corporations, agencies, teams, and other types of groups do not necessarily share a common attitude toward risk, have a cohesive risk culture, or have the experiences to fully understand the larger organizational implications of their individual decisions. People with different attitudes toward risk are not necessarily incentivized to work together to uphold some risk culture mandated by others. Just like the children in the playground, we are much more motivated by our individual sense of risk and safety than by the instructions given to us.

There are also psychological factors at play. In some cases, individuals may characterize themselves as being independent of others. In other cases,

DOI: 10.4324/9781003329220-13

individuals may characterize themselves as being interdependent with others. There is the sense of "I" versus the sense of "we." In a risk setting, the behaviors of individuals with a strong sense of independence may not recognize their role in managing their overall community risk compared to those individuals with a strong sense of "we."

In many cases, the risk programs, policies, and practices are developed around an organization's "big picture" view. The big picture sees the organization using overall statistics and views various activities as being interdependent with others. All individuals are components of the larger "we." However, individuals are compensated and rewarded on an individual basis. For example, noncompliance with risk policies could impact individual salaries, well-being, bonuses, and recognition.

The current risk culture in an organization could be good, bad, or anything in between. Regardless of whether people in a group work well together or not, there is an opportunity to develop and build a cohesive risk culture. A core issue in developing a risk culture is understanding the differences between "I" versus "we." Individualism is innate, but a collective sense of "we" is necessary. We recognize that these factors impact the organization and branch far beyond risk and reflect on overall organizational management and leadership.

A major component of organizational management and leadership is the individual and workplace dynamics. For example, many behaviors can seep into a risk culture, potentially diminishing the effectiveness of any risk program, policy, or practice. These include:

> *Demands for incentives:* Individuals may be unwilling to take ownership of new initiatives if they are not directly rewarded for doing so. These individuals may be part of a workplace culture in which it is the norm to do the bare minimum. In a risk setting, this type of behavior can manifest as individual unwillingness to take responsibility for risk initiatives, performing only cursory work on risk activities, or not pitching in to help when risk events happen. Motivating individuals to develop their own sense of purpose and dedication to the larger mission remains difficult. It remains a challenge for leadership to promote this sense of purpose and hire individuals with the same commitment toward the shared purpose.
>
> *Culture of blame:* Problem-solving is a core concept attempted to be taught across schooling at all levels. However, it remains one of the most challenging activities in a group setting. When something goes wrong, such as a risk event or organizational mishap, individuals and groups may focus more on passing blame to particular parties than on solving the problem at hand. This "us" versus "them" mentality may emerge in siloed settings where individuals and groups may not share ownership or have a sense of community with others. Leaders may also be responsible for promoting a culture of blame, mainly if they also focus on the blame instead of converting learnings onto real organizational changes. Human errors and mistakes happen. Kids learn from an early age, "It's OK,

mistakes help you learn." Somehow, as we grow up, we lose the sense of growth that emerges from accidents and mistakes. In a risk setting, this type of blaming behavior could result in individuals responding to risk events with a focus on blame instead of solving the problems and addressing the factors that resulted in the risk event.

Micromanagement: Managers are not necessarily experts on their employees' jobs. An effective leader does not need to be an expert on all facets of an organization. Instead, they have to trust their teams. Effective management focuses on finding the right people for particular tasks and then trusting those individuals to innovate when opportunities arise. Micromanaging feeds into a culture of compliance and hinders innovation.

Distrust: A lack of trust among colleagues, departments, and hierarchies can be destructive. However, a sense of critical trust, in which individuals have a healthy sense of skepticism, is also important. Trust can also be difficult to attain, as it builds over time and across interactions. From a risk perspective, distrust can further challenge efforts to promote community-encompassing risk initiatives. Leaders can encourage trust-building by modeling trust through trusting their employees. They can also promote trust by modeling inclusivity and respect and developing rewards and incentives that encourage trust.

Understanding: Efforts toward risk programs and activities often are founded on good intentions. However, when those programs and activities are developed without input from those who are most closely impacted, the effectiveness of the risk program and activity is at stake. A classic example involves the use of personal protective equipment (PPE). The mandates and purchases of PPE can happen across various areas of an organization, with or without input from the workers who are expected to wear the PPE. Issues with workers not wearing their PPE, or not properly wearing their PPE, are pervasive. Some common reasons why those workers do not meet their PPE policies include:

- PPE hinders job functions: PPE can get in the way, make work processes even more unsafe, or slow down work. This is particularly bothersome in environments where workers are also incentivized to work quickly (e.g., productivity targets and bonuses)
- PPE is uncomfortable or the wrong size: PPE may not be available in various sizes or levels of comfort. However, user testing before beginning a PPE policy may be effective in finding PPE solutions that will be used.
- PPE is not socially accepted: In cases where PPE is optional, there may be social dynamics at play when workers decide whether or not to wear the PPE.

Time-centeredness: Because there is a significant investment in risk practices, stakeholders may question the necessity for the additional workload associated with those practices. They may not see an immediate benefit to the additional workload. They might not ever see a benefit to the additional workload. Paychecks arrive on a weekly basis, while rewards for a good risk program may emerge over years or decades.

In addition to the general workplace issue described here, team or individual dynamics can be harmful to a group and can seep into an ineffective risk culture. For example, these behaviors include:

- *Power politics or "cliques"*: Individuals may form power alliances and leverage those alliances in ways that impact the risk culture.
- *Workplace bullying*: Individuals can abuse their power, retaliate against employees who speak up about risk-related concerns, or may participate in forms of workplace harassment
- *Complacency*: Individuals or teams may be content with the as-is operations of the organization. They may be too busy or distracted to recognize opportunities for improvement.
- *Lack of follow-through*: Individuals and teams may agree to take responsibility for various initiatives, but may not fulfill their promises or may not satisfactorily fulfill those promises.
- *Nay-sayers*: Individuals may be overly focused on the negative aspects of new ideas and initiatives. Instead of problem-solving of finding innovative solutions, those individuals may focus on only those negative aspects, then use those factors to attempt to veto any proposed initiative.
- *Fighters of change*: Individuals or groups may not see value in change or innovation.

There is not necessarily an all-encompassing solution to these general workplace issues described here. Of course, efforts to hire the right people who promote the best culture possible is an excellent place to start. Many organizations use workplace culture surveys that can gauge the current environment. Culture consultants or organizational change consultants can help guide organizations toward improvements. Just as with any other risk initiative, organizational change takes time, can involve handling criticism, and require extensive and potentially uncomfortable efforts to make meaningful change.

Another issue is that the benefits of a strong risk culture often go unnoticed by stakeholders, including management, employees, suppliers, and customers. It is difficult to track the effectiveness of a risk program because risk successes are often invisible, as those successes consist of the absence of risk events or the reduction of the impact of risk events. That is in strong contrast to risk failures, which are highly visible.

Some examples of unnoticed successes of risk programs include:

- Maintaining and resuming operations following a natural disaster because the organization had a pristine disaster plan

- Avoiding a public relations crisis following an operational problem because they had a public relations team that had strategized around this type of event
- Preventing a cyber-security attack because the corporation did an exceptional job with training their employees

Also, because the successes of risk programs can be unseen, even despite significant investment in creating a risk program, leaders may feel less urgency for maintaining and improving the risk program.

Consider the case of an organization with an established risk program. Suppose that organization neglects to maintain the program or provide the necessary resources to improve the program. As a result, the organization is left to invest in improvements to the risk culture only after some type of preventable, but not covered in the risk program, risk event has happened. For example, consider the scenario of a large manufacturing company. The public learned about their practices of sharing misleading information with government agencies and customers and eventually paid large fines and legal ramifications. As a result, the company invested heavily in reorganization. While they already had a risk program, they realized their current policies were not encompassing enough to prevent or mitigate the most recent risk event. The risk program needed to more carefully consider events that happen internally, such as by giving employees a platform to speak up without penalty when they see evidence of risk-related issues. While it's not clear whether these types of policies would really work, it is evident that a truly effective risk program involves a significant cultural investment, which then feeds into more standardized policies and procedures.

Organizations can get around the silence of an effective risk program by having more conversations about risk. There is an opportunity to leverage insights from risk events that did not happen through the study of near-misses. There is also an opportunity to incorporate reporting of near misses and entirely avoided risk events into more extensive conversations about risk and organizational performance.

Leaders who are intent on developing a strong risk culture often receive a lot of questions. Here, we present a list of the most frequently asked questions and some ideas that can help clarify concerns encountered by any stakeholder on the cusp of developing an effective risk culture and risk program.

Q1: Talking about risk only scares us. We would rather focus on good aspects of the organization and create a positive work environment.

Talking about risk can indeed be uncomfortable at times. Thinking about historical risk events can be unnerving and bring up past controversies and stresses. It can be equally troublesome to think about what *could* happen, especially in comparison to events discussed in the news and recent history.

However, there is also a sense of relief that can come with an operational stance on risk and plans for addressing the highest priority risks. With a solid risk plan and culture, there could be less staying up at night worrying about the next risk event and more comfort in knowing that your organization has the tools to address risk in relation to potential risk events.

There is also value in recognizing that a solid risk culture seeps into other elements of an organization in many ways, including:

- Stronger control over operations, which can improve organizational performance
- A more shared sense of mission, which can improve the overall workplace culture
- More efforts toward long-term planning, which can reduce future needs for "putting out fires," which implies addressing nearer-term high-priority problems
- Recognition of limitations or ineffective management policies and incentives, which can lead to real organizational improvements
- More inclusive strategic planning that incorporates input from many stakeholders, which can lead to more realistic goals and a shared sense of ownership for organizational performance

There is recognition that conversations about risk can be very uncomfortable. It would be much easier not to bring up these issues in conversations and decision-making settings. Sometimes blissful unawareness or the ostrich effect of avoiding unpleasant topics can help us cope. However, there are also broader implications for having conversations about these risks:

- Addressing the unspoken issues: Talking about these topics can help address misinformation or rumors surrounding the issue.
- Validate, acknowledge, and address fears related to the risk topic: Talking about these topics can help normalize talking about them and, consequently, allow for more robust risk-related decisions to emerge.
- Identify strategies around addressing the risk: More conversations can help the process to obtain credible knowledge and decision-making related to the risk.
- Mentally cope with the idea of the risk in meaningful ways: One way of overcoming the biases discussed in previous chapters is to have more conversations about risk.
- Normalize speaking up about risk. Hindsight bias is pervasive following risk events. After a risk event, such as following acts of domestic terrorism, news headlines often allude to community members identifying signals or pre-coursers of the risk event before it occurred. If those community members had reported those signals of a potential risk attack, tragedy could have been averted. The issue is with social dynamics – if someone reports a potential risk event, such as by acting as a whistleblower, and they are

wrong, that imposes risk on the reporter. If someone reports the potential risk event and is right, potential retaliation could also exist. The more we normalize talking about risk and normalize taking precautions when risk signals exist, the more normalized it is to report suspicious activity or any other red flags.

Q2: Talking about risk doesn't impact the outcome if a risk event were to happen. We might as well do nothing and deal with the risk event when it happens.

There is some truth in saying that even with a pristine risk program in place, risk events can and do happen. The discussion in Chapter 3, relating to black swans, perfect storms, and other risk events that may not be easily foreseen, reminds us that we cannot predict or avoid every single risk event that can happen. Thus, we cannot necessarily manage risk for every single risk event that can happen. However, that is not the whole story.

A solid risk program not only addresses the avoidance of risk events but also resilience, such that an organization can effectively and quickly recover following a risk event, regardless of the risk event. In effect, the concept of resilience is agnostic to the type of risk event that occurred. While we can't predict or protect ourselves from every risk event, we can take steps to improve the recovery after the event occurs.

In addition, while we cannot avoid every single risk event, we have options to manage the risk. We can invest in things like:

- Reducing the likelihood of the event
- Reducing the impact of the event
- Transferring the risk to someone else, such as by insurance
- Exiting the activity that opens up the possibility of the risk event happening

Q3: We don't have the money or time to address risk.

A solid risk program does not necessarily need to be expensive or highly time-consuming. We can say the opposite is also not true. Recent history has shown many examples of large firms with extensive risk programs also being largely impacted by risk events. In retrospect, those risk events could have been or were addressed in the risk program.

A risk program needs to be supplemented by a strong risk culture. Otherwise, the program is just words on, likely, many pieces of paper. A solid risk culture relies heavily on aspects that are often "free" or innate to an organization, such as:

- Having a shared sense of mission
- Being open to innovation in ways that can address risk
- Being open to discussing risk-related concerns that arise

- Giving stakeholders the ability to speak up if they see evidence of risk-related problems
- Having humility in recognizing that no organization is immune to risk

It has been argued that these elements of a solid risk culture are more easily formed in organizations that are not necessarily resource-rich. While developing a risk program takes time and energy, it is a relatively small investment compared to the time and energy spent on communication and implementation. However, once a risk program is in place and an appropriate risk culture is formed, the risk practices can be embedded into the existing organizational management. Shared risk culture is not meant to be burdensome but instead, an additional facet or management that is already in place.

The end argument is also that a solid risk program can open avenues for cost and time savings. For example, less time may be needed for "putting out fires" in dealing with smaller and more urgent risk issues, particularly if the underlying factors influencing those issues are addressed in the new risk program. The costs of creating and implementing a risk program are likely to be minor in comparison to the costs associated with lawsuits, fines, and other losses that can occur if a preventable or addressable risk event were to occur.

Q4: Since no major risk events have happened to us before, we don't need to worry about risk.

You and your organization may be incredibly fortunate. Good fortune should always be celebrated. However, not encountering a risk event before doesn't necessarily imply that you and your organization have not had a risk event or are immune to a future risk event.

The first question is: Are you sure that no major risk events have happened before? The collective "us" of individuals and organizations is sometimes very narrowly defined. Often, "us" is defined as an organization's individuals, shareholders, or decision-makers. Suppose those stakeholders are focused on financial risks, and have not seen a history of major financial risks. In that case, it is understandable for them to conclude that there is no history of major risk events. However, there are many types of stakeholders of an individual or organization. Those stakeholders may see more comprehensive history of risk events, such as those related to worker safety, environmental degradation, reputation, and many other types of risk.

Also, there is a question about defining a "major" risk event. What may not seem "major" to some stakeholders may indeed be "major" for others. For example, workplace safety incidents might not look "major" on paper or in safety reporting, but have an enormous impact on individual employees.

One essential element of a risk program is the collection of information about risk events that occur. Major risk events may not be recorded, and consequently, be assumed to not exist. It may be relatively easier for organizations with no current risk program to not see any history of major risk events.

There may also be underlying or hidden risk issues that are unseen by decision-makers and leadership. For example, social or management dynamics may encourage workers not to report injuries. Leaders may only hear what others want them to hear and only hear positive aspects of organizational performance.

Some of the precursors to future "major" risk events may be known, but they are not seen as precursors. For example, there may be a history of narrowly averted risk events. Suppose a manufacturing facility has smoke alarms that frequently malfunction. The smoke alarms may sound a warning when no fire danger is present or may not work when fire danger is present. While the malfunctioning smoke alarms may not yet have led to a major fire in recorded history, the potential for a future fire remains. It's imperative to address the precursors early before larger problems arise.

Q5: We have a risk manager. Risk is someone else's job. I have to focus on my own work and let the risk manager do their job.

If you and your organization have a risk manager, that risk manager should actively remind everyone that risk is *everyone's* job.

Often, organizations have a lot of risk policies that only get discussed at a managerial level. However, effective and sustainable risk policies are informed by perspectives across an organization.

Q6: We don't need a risk program. We put work into our existing risk program for years and no risk events happened. We wasted our time.

Here is one of the most significant concerns with effective risk management: A successful risk program is often silent and unnoticeable by many people. The absence of risk events is unnoticeable, while the occurrence of a risk event is widely apparent.

One way to promote a collective culture around risk is to frequently report and have conversations about the program's effectiveness. Most commonly used risk programs (ISO 31000, Enterprise Risk Management) call for regular reporting and maintenance of risk policies. Reporting can attempt to cover aspects of averted risk events that can be attributed to a well-functioning risk program.

An individual or organization that doesn't see value in a risk program may also struggle to implement activities within the risk program. There may be an opportunity to stop and reconsider why they have a risk program to begin and decide whether the risk program needs to undergo a thorough review.

Q7: I abide by all of the relevant safety regulations and standards. I don't need to think about risk.

Some risk-related regulations and standards can be very politicized depending on the context. These programs may be developed using values that do not necessarily align with your values or those held in your organization. You may see the need for more stringent requirements and standards, which you have the option to develop internally. Also, these risk-related regulations and standards are often developed in reaction to risk events. For example, policies for the use of seat belts in vehicles only developed after actualizing that a problem of vehicle safety existed and that an effective and relatively low-cost solution was available to be mass-produced and enforced. The topics of black swans and perfect storms can remind us that not all risk events can easily be foreseen. Thus, existing policies and standards may not be sufficient to address the new risks on the horizon.

We are also prone to biases that cause us to easily be less aware of our own risk exposure. Consider the single-action bias, in which we assume that if we make a single action to address a particular risk, we have done our due diligence in addressing the risk. However, that single action may, in reality, do little for risk reduction. For example, one can assume that if they take their own grocery bag to the grocery store, they have done their due diligence in addressing environmental risks. However, they may neglect to view their purchases of foods with plastic packaging as equally environmentally harmful. Similarly, meeting risk-related regulations and standards may prompt us to assume that we are taking the appropriate actions, but in reality, there is more to be done to address risk.

Q8: It's better if I handle risk in the ways prescribed by those I trust (positions of authority, elected leaders).

We recognize that risk decisions are based on both data/information/analysis and values. While we may gravitate toward trusting leaders who share our values, there is no guarantee that we fully agree with those values. We may also not be aware of the data, knowledge, and assumptions others use when they develop their risk stances.

We also should be mindful of the motivations of those who attempt to prescribe the risk-based actions of others. Some may be incentivized to deceive for various purposes, such as financial, political, or reputational rewards. Some may have no incentive other than for public safety.

Q9: If I'm not aware of a risk, I don't need to do anything about it.

Let's first distinguish between being subject to a particular risk and being aware of that risk. The term risk *exposure* represents how someone, something, or some organization is subject to a risk source. For example, we are *exposed* to risk when entering vehicles and driving on roadways. The more we drive, the more *exposed* we are to the risk of a car accident. As another example, consider the risk associated with smoking cigarettes. The more we smoke, the more exposed we are to the health effects of smoking, such as lung cancers and other health problems. We can be *exposed* to risk without being *aware* of the risk.

As discussed in previous chapters, our sphere of awareness of risk issues is limited. Our awareness of risk is largely based on our own experiences, the media (news or social media), our daily conversations, and any other communications with other parties.

Consider first the information encountered in the news. Those news headlines we see contain topics deemed to be newsworthy by the news outlets we use and the algorithms that attract us to various news outlets and articles. The newsworthiness of a topic does not directly translate into high-priority risks, but it could. Here are some qualities of a news article that make it newsworthy:

- *Timeliness:* Information about recent events or current information is interesting for readers because it provides new information that could be surprising or particularly interesting. Some of the most persistent risk issues are not necessarily new or newsworthy enough to be on the front page of a newspaper, such as risk related to smoking, accidents, and chronic diseases.
- *Proximity:* Stories about topics that directly impact readers, such as information from the local community, or activities the readers engage in, are newsworthy. The proximity of a news article would vary depending on the type of news source and the target audience. Due to this issue of proximity, we may not be aware of risk issues that impact other stakeholders. Additionally, we may not be fully aware of how the risk issues impacting other stakeholders may indirectly impact us. For example, during the COVID-19 pandemic, COVID-19 rates in individual communities had a larger global impact, through the spread of COVID, the impact on workforces, travel policies, and the availability of goods and services. As discussed in previous chapters, effective use of risk science should contain an understanding of the various stakeholders of a system.
- *Controversy:* Readers are attracted to stories that have conflict or controversy. There could be many subconscious, societal, and social reasons for this, which we will not discuss here. Those controversial topics on risk-related issues that we encounter in the news may not necessarily be the most relevant topics for our lives. Those controversial topics could be rare, poorly understood, or may lack the detail needed for us to understand if they are relevant to us. Some major risk-related topics have little controversy involved, such as seat belts, obeying traffic signals, and food safety, so we may not frequently encounter those topics in the news. Conversely, highly controversial topics may have more extensive coverage in the news, including gun control, vaccines, and healthcare, which all also have significant risk-related implications. Of course, highly controversial

topics can also be politically divisive, making it even more difficult to form balanced and objective stances on those issues.

- *Human interest:* Readers are attracted to stories about other people. Again, there may be many subconscious, societal, or social reasons to explain this, but we won't speculate about those reasons here. Some examples of human interest are stories about kindness, enduring extreme circumstances, or significant accomplishments. Readers can identify with those stories and then feel a sense of motivation or have positive thoughts. Human interest stories may help with implementing a risk science approach because it encourages readers to understand various stakeholders and consider the broader social impact of their decisions.
- *Relevance/Impact:* Readers seek information that can help them understand relevant topics and to help them make related decisions. Business news outlets often focus on business-related issues, such as related to financial trade, pressing regulatory issues, or growing trends in management and decision-making. Encountering relevant news can help us manage risk in our lives and organizations. However, if we only get news from business news outlets, we are limiting ourselves to only business-related risk issues.
- *Prominence:* News outlets often cover information sourced from highly visible celebrities, political figures, and other popular and interesting individuals. Those popular individuals may use their visibility and relatability as a platform to speak about risk-related issues. High visibility does not necessarily guarantee that the individual is an expert on risk-related issues, is informed using credible information, or shares risk-related values with the readers.
- *Novelty:* Some news outlets may promote articles that contain unusual or novel information or insight. From a risk perspective, these articles contain information about unlikely risk issues. This may alter our sense of likelihood for particular risk events. For example, suppose a news article that discusses a meteor hitting Earth and causing extensive damage. While this event is certainly possible, is recorded in history, and could impact individual readers, it may not be a highly likely event for most readers.

Additionally, the headline or title associated with a news article may not match the content. Terms like "clickbait" describe how news outlets can alter the general theme of the article in an attempt to get readers to click on the article. Also, readers may have a propensity to only read those headlines without actually reading the information in the article. All of these issues can limit and alter our awareness of risk-related issues.

The newsworthiness issues also translate into further limiting or altering our awareness of risk-related information through other forms of communication. Examples include:

- Social media posts are more engaging through the use of images, graphics, and videos, are simple, prompt the viewer to act in some way, and are timed to be visible when target audiences are using the service.

- Internet searches: Internet search engines use algorithms to determine the best or highest quality matches for our Internet searches. Some factors used to determine relevancy for a search result could include collected information about the user (e.g., inferences from past searches, indirect information about values/beliefs), location, language, and type of device. Also, content creators can use search engine optimization to allow their website to be more visible in a web search.
- Marketing/promotional materials: Effective promotional campaigns, used by businesses, political candidates, or other organizations, have narrowly defined goals, a well-defined target audience, and messages that are catered to that target audience. These campaigns may also use many different marketing channels and technologies that allow the core message to follow the user across applications (e.g., email, social media, mail, and apps)
- Ordinary conversations: Interesting conversations often include topics that are common among the conversation participants, topics that are important to participants, personal opinions, and topics encountered in the news. Controversial topics are often avoided, but when they are encountered, they can be handled delicately. As a result, when points of disagreement around risk-related issues are encountered, the disagreement can be handled in such a way to avoid conflict.
- Signage and risk-related messaging: Individuals see messages that attract attention. Some classic examples include flashing warning signs, alarms, and other signals generally used to promote safety.

The communication strategies discussed here suggest that individuals often encounter oversimplified and targeted information about highly complex risk topics. These individuals may not be aware of how limited their awareness is for risk-related issues, or whether their awareness has been steered in a particular direction through media and other social mechanisms.

While individuals may consciously try to open up their spheres, they have limited capacity for awareness. Of course, it's unrealistic to think anyone has the time, energy, or interest to be aware of every topic and every perspective on that topic. However, there is always an opportunity to clear the pathway to understand risk using a practical approach that is founded in science.

Q10: If I address my own personal risk, my problems will be solved.

Risk does not only impact us as individuals, but there are many societal risks to consider.

We make a lot of decisions that impact societal risk. For example, these include:

- Electing leaders who are responsible for policymaking that can impact large risk issues, like national defense, management of our critical infrastructures, environmental management, and healthcare policy.

- Taking a risk science approach to management decisions in our professions. For example, if your organization has poor or unestablished risk policies, there could be potential major risk-related issues.
- Adapting our behavior to address risk issues: Recognizing our role in understanding and managing risk at the community level is critical. Consider the case of the COVID-19 pandemic, in which all individuals had an essential role in managing the community and the global risk.

Thinking about risk is only a starting point. It is a job that is never complete because new risks will always emerge. Rules and regulations are constantly adapting and changing. Of course, while some of these rules and regulations can improve our ways of understanding and managing risk, this is not necessarily guaranteed. Also, technologies are constantly innovating, allowing us to adapt and find new ways to address risk. As a result, we can always find better ways of addressing various risks.

Let's consider the example of regulations related to vehicle seat belts. The movement toward promoting seat belt use was a major achievement in risk. However, this movement was just a start. Seat belt technologies continue to develop. There are also many criticisms of current seat belt technologies and vehicle safety in general, as crash test dummies used for vehicle safety testing may not be representative of the characteristics of all occupants of vehicles.

As another example, the topic of continuous improvement to address risk is crucial for addressing workplace safety risk. Testing personal protective equipment has shown that standard equipment is not necessarily effective for all individuals and in all scenarios. As an example, the first space suits were designed for all-male crews. Those same space suits on women may pose serious safety concerns. Because poorly fitting PPE is a significant cause of injury across professions, it is critical to continuously improve testing and understanding of this PPE across scenarios, uses, and individuals.

In summary, the job of thinking about risk is never complete. However, the more risk science is infused into a risk culture, the easier it is to create and maintain an effective risk program.

Key Takeaways

- A risk culture consists of shared beliefs, norms, values, practices, and structures with respect to risk in an organization.
- Corporations, agencies, teams, and other types of groups do not necessarily share a common attitude toward risk or have a cohesive risk culture, but managerial practices can incentivize adherence to risk standards and initiatives.
- Following basic risk standards and regulations may not be sufficient for overall risk management.

11 To manage risk is for our own benefit

Learning Objectives

After reading this chapter, you will develop further knowledge on aspects of:

• Benefits of active risk management

Many business executives share stories about the stresses associated with worrying about untreated risk in their individual lives and organizations. In some cases, stress can be helpful. The fight-or-flight response can be essential in handling emergencies. From a risk perspective, some stress can be helpful, particularly if that stress enables individuals and organizations to act in a way to appropriately address the sources of those stressors. As a result, individuals learn from those stressors and can incorporate those learnings into more advanced risk-related processes. Other types of stress can be harmful, as they can lead to poor outcomes in the form of destructive behaviors, unhealthy coping mechanisms, or activities that can increase the potential for risk events.

Let's first discuss the positive implications of risk-related stresses. Treated and understood risk can lead to individuals and organizations investing in risk science, risk preparedness, continuity plans in cases of a risk event, and recovery plans. This type of preparedness and planning can have many benefits, including:

• Encourage the development of risk standards and policies
• Improve operations, resulting from comprehensive studies of risk in operations
• Improve and establish a risk culture, including overall awareness and training for the risk-related issue
• Ability to consider and study risks that may not have been apparent before performing a comprehensive overview of enterprise risk
• Reduce potential liabilities in cases of risk events
• Reduce risk-related incidents

DOI: 10.4324/9781003329220-14

Viewing risk as something that can be understood and treated results in improved organizational awareness and management. However, chronic or bad stress can be destructive for both the individual and the organization.

First, there are psychological and work-performance symptoms of stress. For example, studies have examined relationships between chronic stress and heart disease, high blood pressure, diabetes, anxiety, depression, poor job performance, burnout, absenteeism, poor job satisfaction, and job turnover.

Stress can also be associated with physical symptoms. For example, studies have examined relationships among stress, injuries, and risk-taking behaviors. As a result, risk can beget additional risk.

Individuals employ various activities and behaviors to cope with stress. Individuals can change or adapt how they think, in the form of denial or changing their perspective in a way that distances them from the issue. In a risk setting, individuals may prefer, even in a knowingly false way, to deny the possibility of high-priority risk events. While that is a personal choice to do so, this type of denial can impact stakeholders.

Individuals may instead choose to use their behavior to cope with the stress. They may work on problem-solving to address the source of the stressor. Some behaviors can be productive, such as physical exercise, recreation, mindfulness, and seeking support from their network (family, friends, workplace, mentors, and professionals). Other coping mechanisms may allow the individual to escape from the stress. These strategies could result in poor health outcomes related to smoking and alcohol. Again, employing these types of escapist coping behaviors can lead to additional compounded risk for the individuals involved.

Increased stress can impact the workplace and team performance in an organizational setting. Studies have examined how stress can affect the ability of teams to form consensus, make decisions, share responsibilities, and handle conflict. This poor workplace culture can lead to inefficiencies and low productivity, leading to long hours, unbalanced workloads, job insecurity, and poor morale. Overall, this can result in poor quality of work and reduced ability to handle risk events when they occur.

A risk science approach can result in currently unknown risks being more visible. Some may argue that knowing about additional risks can increase overall stress. That may indeed be a short-term implication of using a risk science approach.

The risk science approach brings attention to the fact that risk events are not necessarily foreseen. That in itself can also contribute to overall stress. However, the risk science process also includes concepts of resilience, allowing individuals and organizations to develop strategies to maintain functionality and recover following any type of risk event.

Awareness of risk issues, or lack thereof, is only one component of risk science. The approach includes considering knowledge, analyzing risk, and deciding how to address those risks. Those decisions for investments and activities to address the risk happens with input from those stakeholders who may share the stress burden. Also, addressing the risk is a shared responsibility, and some individuals and units have accountability for their related responsibilities. Thus, when a risk science approach is used, the stress is shared and addressed using systematic standards and policies.

In summary, while more knowledge about risk can lead to additional stress, there can also be benefits to leveraging that knowledge to manage risk. Consider the famous Fred Rogers' quote: "When I was a boy and I would see scary things in the news, my mother would say to me, 'Look for the helpers. You will always find people who are helping.'" This is true for many risk contexts widely discussed and shared in news headlines. For example, during the COVID-19 pandemic, there was constant media attention on health risk. However, these headlines were also supplemented by scientists, healthcare workers, and individual citizens providing assistance and innovations to help address this risk event.

Key Takeaways

- Treated and understood risk can promote risk preparedness, continuity plans in cases of a risk event, and recovery plans.
- Untreated risk can lead to stress, and increased stress can impact overall workplace and team performance.

12 Piecing it all together

Examples of the risk science process

Learning Objectives

After reading this chapter, you will develop further knowledge on aspects of:

- Examples of how the concepts of this book can be applied in a real setting
- Challenges that can be encountered when applying the concepts of this book

---

Let's first review the steps in the risk science process in which one uses applied data. We will break it down into numbered steps that align with risk study components described in Chapter 5.

Hypothesis and study design: This step consists of asking a question to be answered by the risk science process. This could relate to understanding the level of risk associated with some single activity (e.g., the potential for a severe weather event) or relate to overall risk processes (e.g., enterprise-level risk).

Data and information: In this step, we collect data and information that will be useful for addressing the overall risk-related question. We seek high-integrity data and information, but we also recognize that there are limitations. Thus, we understand that we may have low knowledge associated with the question and that low knowledge should be considered throughout the risk science process. We also can increase knowledge in many ways, such as through data collection, expert elicitation, and scientific research.

Analysis: Where there are nuances in a professional setting, we can generalize that Analysis can be used alongside the terms "Risk Analysis" and "Risk Assessment." Generally, this step consists of the process of understanding the nature of the risk while using the available knowledge, recognizing that knowledge can also be increased or improved during this process.

DOI: 10.4324/9781003329220-15

Management review and judgment: This step consists of interpreting the analysis. This includes summarizing, interpreting, and deliberating the results of risk assessments and other assessments, as well as of other relevant issues (not covered by the assessments), to make a decision.

Decisions and communications: Finally, this step consist of making a decision to determine how to address the risk. As discussed in Chapter 4, the options generally are:

- Acceptance
- Transfer
- Mitigation
- Avoidance

Deciding on the appropriate activities or investments involved with those options is called decision-making.

The communication process may begin before or after the decision-making component. In some cases, communication can inform and engage with stakeholders to make decisions. In other cases, communication may inform stakeholders about the outcome of decisions. Risk communication involves sharing risk-related data, information, and knowledge among different target groups. These groups can include individuals, businesses, regulators, the media, the general public, and other stakeholders.

We will follow the steps outlined here to discuss the two examples in this chapter.

EXAMPLE 1: Prediction of a severe weather event

Let's assume that we are only observers of the risk science process. The risk related to a severe weather event is not limited to a single stakeholder. But instead, the responsibilities are held across various entities. We have scientists with weather forecasting supercomputers that analyze current weather conditions and make predictions for future weather events. We have journalists and meteorologists who interpret the information from those supercomputers and offer the best judgments or interpretative guidance in the form of weather predictions and, potentially, other risk-related communications for emergency preparedness. We have businesses, agencies, and individuals who interpret those weather predictions. Finally, all individuals and organizations act based on their perception of available risk-related information.

We can approach this process from a high-level perspective. In general, the steps consist of:

Hypothesis and study design: The question to be answered in this process is: How can we characterize the risk of a major hurricane approaching a highly populated area in the next three days?

Data and information: To understand current weather conditions, significant amounts of data are collected. Observations are collected from many sources, including the following:

- Weather balloons measure atmospheric pressure, temperatures, wind speed, and so forth.
- Buoys collect data including wind speed, sea temperature, wave energy, and so on from coastal and offshore waters.
- Satellites collect information on visible weather patterns, cloud cover, surface and water temperatures, and so on.

Analysis: The observations collected in the previous step are then aggregated and processed using supercomputers. These supercomputers run weather-related activity models, which are then used to form forecasts. Several entities form their predictions about a storm's path. A handful of global forecasting models can be combined to form consensus models, each with varying levels of historical accuracy. These models are also slightly different, running on different sets of assumptions, data, and model designs.

Suppose the analysis results are simplified to assume a 40% likelihood of a severe hurricane impacting a highly populated area. However, this is with a very low knowledge strength, as these various models used to predict the path of the hurricane show widely different results. Thus, while the likelihood is medium at 40%, the knowledge is low. The analysts, scientists, and meteorologists tracking this hurricane have performed their assessment and judgment that is communicated to the local community. They suggest that all residents take the necessary precautions for storm preparedness, including preparing emergency supplies.

Management review and judgment: The interpretation of the analysis and the results can be challenging, as it involves likelihood judgments that are not easily understood. There are uncertainties with the model prediction. The models do not all agree with one another. The accuracy of these models decreases over time, as a forecast one day out is more predictable than a forecast for three days out. Many individuals and organizations review the analysis results and form their own judgments. These stakeholders include:

- National and local weather service agencies
- Businesses and other organizations
- Individuals

Decisions and communications: Different stakeholders use these results in different ways:

- National and local weather service agencies: Distribute warnings to share the potential for a severe hurricane. However, the agencies make their assumptions about how to take an uncertain forecast with a 40% likelihood and decide how and where to issue warnings.

- Businesses and other organizations: Determine how to respond to this 40% likelihood of a severe hurricane. They may adapt their operations, such as re-routing freight transportation routes, pausing operations in particular areas, encouraging employees to evacuate, or mobilizing relief services.
- Individuals: Determine the most appropriate steps to prepare individually for a storm. With a 40% likelihood for a severe hurricane, some individuals may not take the storm seriously and choose not to act. Others may decide to evacuate. Others may choose (or be required to) to stay while equipped with storm preparedness supplies.

National and local weather service agencies employ experts in risk communication activities. They study the most effective mediums, technologies, words, colors, sounds, and formats for sharing emergency warnings. These agencies also work with emergency managers, businesses, government officials, and the general public to develop plans for weather readiness.

Businesses and other organizations likely also have emergency communication plans. For example, many businesses have emergency alert systems in place. They also have internal communication processes in place for sharing changes in operational activities in reaction to potential weather events.

Individuals may also have informal communication plans in place. For example, they may seek to talk to family and friends to share a warning and provide suggestions for next steps.

In this example, the risk process is owned by many stakeholders in society. Across each step of the risk science process, some individuals and organizations have specialized responsibilities. There is also a recognition of uncertainty – the likelihood of a severe hurricane was 40%, and that likelihood was based on weak knowledge. However, a severe weather event can have irreversible damage with health and safety implications. Therefore, the precautionary principle, as discussed in Chapter 1, which applies when there are large scientific uncertainties, can guide these stakeholders to act in a way to protect things of value (health, safety) in the face of those uncertainties.

EXAMPLE 2: Enterprise-level risk management for an agricultural facility

Suppose you manage a local farm. Your mission is to improve access to healthy foods within your local community. Your organization owns 10 acres of land that contains both greenhouses and open farmland. Community partners and volunteers enable your organization to produce food distributed to local community members in need.

Your organization is largely mission-driven and not profit-driven, as it relies on funding from local businesses and individuals. Because so many community members in need depend on your organization, you feel it's imperative to consider risk at the enterprise level, meaning that you would like to think about how to address all the highest-priority risks that are facing your organization.

In a strategic planning meeting, you ask your board of directors to consider some risk issues that concern them. The directors have many concerns, including:

- Climate change making it more difficult to continue farming operations
- Extreme weather events
- Reduction in volunteers
- Food safety problems
- Drastic increases in operating costs

What surprises you most is that you initially thought your question was simple and would only result in a few concerns from your board of directors. Instead, after they voiced a few concerns, many other concerns emerged. You finally realize that there are many risk issues and concerns. There is a need to consider which risk issues are the highest priority.

You decide instead to use a risk science approach to thinking about risk. The steps are described as follows;

Hypothesis and study design: You and your board of directors would like to decide how they should allocate their resources and attention toward the most important risk concerns. You have risk training and decide to conduct a group exercise to identify potential risk events and to analyze the risk using available data and information found using their expertise. Your group will then reconvene to review the analyses and determine the next steps.

Data and information: First, you lead a meeting with the members of the board to discuss risk events that have occurred in the past, then brainstorm about additional or emerging risk issues. You also develop a survey to gauge risk concerns from the volunteers and the local community.

Analysis: Using the survey results and discussions among the board, you use the structured what-if technique (SWIFT) to identify events that could negatively or positively impact your local farm. Some questions that emerge are:

- What if the farm cannot afford to operate (costs)
- What if the farm can no longer produce food (lack of volunteers, climate change, extreme weather, drought, pests, weeds, fungal diseases)
- What if demand is lost or is too high (customer demand)

Table 12.1 shows the resulting events that are selected for study. You collect data and conduct the analysis. You are qualified to interpret information from credible sources, including government agencies, certification agencies, and

Table 12.1 Risk events and data/information for local farm example

Events	Data/Information
Drought	Weather patterns, and climate change projections conducted by credible scientists and organizations
Pests, weeds, and fungal diseases	Rules and policies for organic certification; data and resources offered by organic certification agencies
Excessive operating costs	Current financial records; news articles about projected changes in costs for things like taxes, water, organic fertilizers, etc.
Climate change	Climate change projections collected from environmental agencies and weather agencies
Extreme weather events	Historical records of extreme weather events, such as hurricanes and tornadoes, which are common in the region
Lack of volunteers	Survey of volunteers to understand how likely they are to continue volunteering

community members. You also supplement the analysis with expert opinions from your network of connections.

For most of the events shown, the analysis consists of compiling information or projections from various sources and forming their own opinion or stance that will be eventually shared with the board. For example, the predictions of drought events are supplemented by climate change projections and weather patterns. You also consult local meteorologists to ensure they use the correct data and properly understand that data. As another example, when seeking to understand the risk related to the lack of volunteers, you collect information directly from volunteers. Because you are experienced with matters of human resources, you understand the nuances and how to interpret the feedback given by the volunteers.

Using the updated data/information from Table 12.2, further analysis characterizes the risk related to the studied events. Table 12.2 summarizes the risk analysis plan and results. The results show that some of the risks of concern are widely studied and apply to many community facilities. In other words, this agricultural facility is not unique in considering these risks and therefore can seek to understand what other organizations are doing to address these risks.

In general, the analysis results suggest that some events, such as drought, seem highly unlikely, though the knowledge supporting this judgment is rather weak. Other events related to pests, weeds, and fungal diseases are prevalent and highly studied by the scientific community and the private sector. Thus, this risk has relatively lower uncertainties due to the nature of the risk and the clear testing and recorded history of pests, weeds, and fungal diseases impacting the local community.

Table 12.2 Risk analysis plan and results for local farm example

Events	Analysis	Analysis Results
Drought	Review data/ information to form a stance or opinion	A drought seems unlikely as this region is not prone to drought activity. The local farm has access to irrigation services that can be used if drought conditions are persistent, assuming local regulations allow.
Pests, weeds, and fungal diseases	Review data/ information to form a stance or opinion	This issue is widespread among all agricultural facilities. Because this farm is certified organic, only allowable treatment options are considered. This remains a high concern based on the analysis results. Therefore, more study is needed to determine the most appropriate treatment option.
Excessive operating costs	Review data/ information to form a stance or opinion	History has shown that the farm is in sound financial condition. The farm is funded by external donors committed to the overall mission. If financial hardships occur, the external donors are willing to increase support.
Climate change	Review data/ information to form a stance or opinion	The results are very unclear. This is a politicized topic, and the time horizon associated with this issue is very far out. The board remains primarily focused on near-term issues, mostly looking a few months to one year in the future.
Extreme weather events	Review data/ information to form a stance or opinion	Extreme weather events are very common in the region. There is need to be aware of weather warnings and have preparedness plans in place.
Lack of volunteers	Trends for volunteer sentiment toward working for the local farm	Issues with lack of volunteers seem currently unlikely. Volunteers are committed to working at the farm and no shortage currently exists.

Management review and judgment

The board of directors then meets to discuss each risk event. They determine the severity, likelihood, strength of knowledge, and risk for each event, as shown in Table 12.3.

However, the board members feel uncomfortable with some likelihood and severity judgments. For example, in reference to climate change, they are unsure about what specific events related to climate change they are considering. The events could include hotter temperatures, more severe storms, drought (already covered separately in the table), poverty, and other events that are not easily simplified in a single table. As a result, it's unclear how to interpret a "Medium" severity or a "Medium" likelihood, as the board asked, "Medium what?"

Table 12.3 Risk characterization for local farm example

Events	Severity	Likelihood	Strength of Knowledge	Comments	Risk
Drought	Medium	Low	Medium	Ensure irrigation systems are operational	Low
Pests, weeds, and fungal diseases	High	High	High	Noted dangers of using pesticides; study of less harmful pesticide solutions	High
Excessive operating costs	Low	Low	Low	Strong donor commitment to help with operating costs	Low
Climate change	Medium	Medium	Medium	Long-term issue, but board is focused on short-term issues	Medium
Extreme weather events	Medium	Low	Low	Commitment to ensuring the safety of volunteers, animals, and other stakeholders	Medium
Lack of volunteers	High	Low	High	Maintain a strong positive relationship with volunteers	Low

The board also recognizes disagreement with the severity and likelihoods presented in the table. For example, with the Drought event, there is only one scenario present: a "Medium" severity and a "Low" likelihood. The board questions whether this is the only possible scenario for this event. For example, what about a "High" or "Low" severity? In future iterations of the study, they should consider additional and more specific scenarios for each risk event.

The board of directors determines that some risks are deemed to be of low priority. For example, issues with climate change are seen as long-term issues, while the current focus area is on very near-term issues. Thus, even though climate change is a large global issue, this farm is committed to not contributing to climate change. The farm remains certified organic and reserves a "do no harm" philosophy.

Other risks are deemed to be of high priority. The issue of pests, weeds, and fungal diseases tops the list of highest concerns among all the board members because they have witnessed the impacts of pests in agricultural facilities. The lost crop due to this risk seems to be wholly avoidable if proper agricultural practices are implemented. They remain committed to being an organic facility, but they also know that there are effective organic-friendly activities that can be used to protect the crop.

The risk of excessive operating costs is a large concern, as it can have serious financial implications. However, the risk is deemed to be low. Because some board members are donors and conduct outreach activities with other donors, the board is confident that the donors will continue to support the organization, even if operating costs rise.

Decisions and communications

The activities conducted earlier in this process enable the board to discuss the questions, including "which risks are acceptable?" Now, the board can systematically examine activities or investments that can be used to address those risks.

Table 12.4 shows the resulting risk management decisions. Many of the risks are deemed to be acceptable, while others require mitigation.

In general, the board had a broad consensus on these risk management decisions. However, they recognize that they will not always agree on every analysis and resulting decision. Therefore, they decide to later develop procedures around their risk and decision-making processes, just in case disagreements emerge.

The board also recognizes that they may have missed some risk issues. For example, in their initial discussions, they discussed food safety issues. Maybe there are other looming risk events that haven't yet been considered. These could include problems with public reputation, pandemics, supplier shortages, and so on. Clearly, the process is incomplete, but this experience served as an important starting point in a long journey of considering and managing risk. What is needed is some additional facilitation and risk expertise for more comprehensive risk assessment and management.

Finally, the board begins the risk communication process. This communication is not meant to be one-way communication. Instead, some issues require stakeholder input and engagement. Other issues require more direct communication mechanisms.

The board needs to develop an emergency preparedness plan to address the risk of extreme weather events. This plan needs to include the ability to communicate with volunteers in the event of a hurricane, tornado, or some other emergency. The board seeks input from the current volunteers and determines

Table 12.4 Risk decisions for local farm example

Events	Risk	Decision	Activity/Initiative
Drought	Low	Acceptance	Do nothing
Pests, weeds, and fungal diseases	High	Mitigation	Conduct additional risk studies to determine best practices, then adopt best practices
Excessive operating costs	Low	Acceptance	Do nothing
Climate change	Medium	Acceptance	Do nothing
Extreme weather events	Medium	Mitigation	Develop emergency preparedness action plans to protect people and animals at the local farm
Lack of volunteers	Low	Acceptance	Do nothing

the most effective ways to share emergency information with those individuals. The board realizes that there is no quick way to send out this type of communication because contact information about volunteers is not centralized or easily accessible. The board also needs to develop a policy to decide when to send volunteers home and cancel planned events. These policies should be made available to all volunteers, and those volunteers should be prepared to follow all instructions. The board decides to create a committee to address these issues.

The communication for risk related to pests, weeds, and fungal diseases is relatively minor. If the farm chooses to change its methods, it will need to ensure that any changes continue to follow the rules for its organic certification. However, stakeholders may need some engagement when determining the most appropriate new methods to adopt. For example, some methods may cause nuisances (e.g., foul smells, noise, and traffic) to the adjacent communities. The board may also choose to publicize any changes to inform the local community. Information about these changes may interest customers, donors, and residents in nearby areas.

Key Takeaways

- A risk science approach is essential for understanding, managing, and communicating risk issues.
- A risk science approach is practical and useful, even without the use of probabilities and technical tools.

13 Conclusion

Over the previous 12 chapters, we have taken a long journey through the most important topics in risk science. The concepts of risk analysis, risk management, risk perception, and risk communication are essential for all of us as we navigate our personal and professional lives. Through this journey, we have applied the book's concepts to real applications, recognizing that in a real organization, risk issues can be complex, messy, and emotional. Risk science helps us make sense of those real issues.

This book bridges the gap between theory and application. The book highlights ideas and principles rather than formal theories, conceptualizations, and methods. The book is practical in promoting action on the reader's part. This book allows readers to learn about risk topics and immediately apply those topics in an individual or organizational setting. The authors' challenge to you is to take the concepts from this book and try them out. Let us know about the process and any challenges you have.

As you apply the concepts of the book, it's important to recognize that risk means different things to different people. However, advances in risk science continuously help to bridge the gap among those various disciplines. We are constantly finding common ground that can be used to improve our abilities to address risk in creative ways that also allow those different fields to learn from one another.

It's also equally important to recognize that having a structured and systematic way of thinking about and managing risk can benefit both individuals and organizations. First, there is the benefit of having plans and actions to address potential risk. Second, and more importantly, a sense of relief comes with translating worry about risk events into action, with clear evidence-based prioritization of risks and accountability for actions to address the risk.

You may seek to learn more about risk. The authors of this book have also authored a more technical book about risk: *Risk Science: An Introduction*. This book is used in college-university level courses and also in private organizations. While *Risk Science: An Introduction* is more mathematical, it can provide additional clarity on the more technical concepts.

DOI: 10.4324/9781003329220-16

Readers seeking more information about risk can also leverage web resources. The Society for Risk Analysis shares many risk education materials. The website SRA.org contains resources including:

• Webinars developed by leaders in risk science research and practice
• Podcasts provide an informal discussion of risk-related topics
• Videos and training materials
• Lists of universities and colleges with risk education programs
• Guidance documents on risk science terminology and principles

Those interested in spearheading risk programs in their organizations can seek to adopt certifications for ISO 31000, Enterprise Risk Management, or IRGC. The principles in these programs do not always follow the best practices described in this book, but they provide a starting point that can be improved upon through increased discussions and the development of internal standards and policies.

Finally, it's important to remember that the adoption of risk science is never finished. The methods, standards, and policies you develop constantly need updating and refinement. That is the nature of risk. However, compared to the start-up challenges of developing a strong risk culture, maintenance of that risk culture does not seem so large, particularly because the momentum is already in place.

The authors hope the text of this book was helpful as you think about risk in your personal and professional lives.

Appendix I

The many definitions of risk

Risk has many definitions that vary by discipline and application. This appendix shows the varying definitions as shown in the SRA Glossary.[1]

We consider a future activity [interpreted in a wide sense to also cover, for example, natural phenomena], for example the operation of a system, and define risk in relation to the consequences (effects, implications) of this activity with respect to something that humans value. The consequences are often seen in relation to some reference values (planned values, objectives, etc.), and the focus is often on negative, undesirable consequences. There is always at least one outcome that is considered as negative or undesirable.

Overall qualitative definitions:

1 Risk is the possibility of an unfortunate occurrence
2 Risk is the potential for realization of unwanted, negative consequences of an event
3 Risk is exposure to a proposition (e.g., the occurrence of a loss) of which one is uncertain
4 Risk is the consequences of the activity and associated uncertainties
5 Risk is uncertainty about and severity of the consequences of an activity with respect to something that humans value
6 Risk is the occurrences of some specified consequences of the activity and associated uncertainties
7 Risk is the deviation from a reference value and associated uncertainties

ISO defines risk as the effect of uncertainty on objectives. It is possible to interpret this definition in different ways; one as a special case of those considered above (e.g., d) or g)), with the consequences seen in relation to the objectives.

Risk metrics/descriptions (examples):

1 The combination of probability and magnitude/severity of consequences
2 The combination of the probability of a hazard occurring and a vulnerability metric given the occurrence of the hazard
3 The triplet (s_i, p_i, c_i), where s_i is the ith scenario, p_i is the probability of that scenario, and c_i is the consequence of the ith scenario, $i = 1, 2, \ldots N$.

4   The triplet (C',Q,K), where C' is some specified consequences, Q a measure of uncertainty associated with C' (typically probability), and K the background knowledge that supports C' and Q (which includes a judgment of the strength of this knowledge)

5   Expected consequences (damage, loss). For example, computed by:

    a   Expected number of fatalities in a period of one year (Potential Loss of Life, PLL) or the expected number of fatalities per 100 million hours of exposure (Fatal Accident Rate, FAR)

    b   P(hazard occurring) x P(exposure of object | hazard occurring) x E[damage | hazard and exposure]
       i.e. the product of the probability of the hazard occurring and the probability that the relevant object is exposed given the hazard, and the expected damage given that the hazard occurs, and the object is exposed (the last term is a vulnerability metric, see Section 1.19).

    c   Expected disutility

6   A possibility distribution for the damage (for example a triangular possibility distribution)

Note

1 SRA (2018). SRA Glossary. www.sra.org/wp-content/uploads/2020/04/SRA-Glossary-FINAL.pdf

Appendix II
Other risk-related definitions

The following section shows definitions for many of the terms presented and discussed throughout this book. These definitions are sourced from the SRA Glossary.

Uncertainty

Overall qualitative definitions:

- For a person or a group of persons, not knowing the true value of a quantity or the future consequences of an activity
- Imperfect or incomplete information/knowledge about a hypothesis, a quantity, or the occurrence of an event

Epistemic uncertainty: As above for the overall qualitative definition of uncertainty and uncertainty metrics/descriptions

Aleatory (stochastic) uncertainty: Variation of quantities in a population of units (commonly represented/described by a probability model)

Probability (likelihood, chance, frequency)

Overall definition: A measure for representing or expressing uncertainty, variation or beliefs, following the rules of probability calculus.

Different types/interpretations:

- Classical probability: The classical interpretation applies only in situations with a finite number of outcomes which are equally likely to occur: The probability of A is equal to the ratio between the number of outcomes resulting in A and the total number of outcomes, i.e.

  $P(A)$ = Number of outcomes resulting in A Total number of outcomes
- Propensity/frequentist probability: A frequentist probability of an event A, denoted $P_f(A)$, is defined as the limiting fraction of times the event A occurs if the situation considered were repeated (hypothetically) an infinite number of times.

- Subjective (judgmental, knowledge-based) probability: Reference to an uncertainty standard: The probability P(A) is the number such that the uncertainty about (degree of belief in) the occurrence of A is considered equivalent by the person assigning the probability, to the uncertainty about (degree of belief in) some standard event, for example drawing at random a red ball from an urn that contains P(A) x 100% red balls.

Exposure

Exposure of something: Being subject to a risk source/agent (for example, exposure of asbestos)

Event, consequences

Event:

- The occurrence or change of a particular set of circumstances such as a system failure, an earthquake, an explosion or an outbreak of a pandemic
- A specified change of the states of the world/affairs

Consequences: The effects of the activity with respect to the values defined (such as human life and health, environment and economic assets), covering the totality of states, events, barriers and outcomes. The consequences are often seen in relation to some reference values (planned values, objectives, etc.), and the focus is often on negative, undesirable consequences.

Harm, damage, adverse consequences, impacts, severity

Harm: Physical or psychological injury or damage
Damage: Loss of something desirable
Adverse consequences: Unfavorable consequences
Impacts: The effects that the consequences have on specified values (such as human life and health, environment and economic assets)
Severity: The magnitude of the damage, harm, etc.

Hazard

A risk source where the potential consequences relate to harm. Hazards could, for example, be associated with energy (e.g., explosion, fire), material (toxic or eco-toxic), biota (pathogens) and information (panic communication).

Knowledge

Two types of knowledge:

Know-how (skill) and know-that of propositional knowledge (justified beliefs). Knowledge is gained through, for example scientific methodology and peer-review, experience and testing.

Opportunity

An element (action, sub-activity, component, system, event, etc.) which alone or in combination with other elements has the potential to give rise to some specified desirable consequences

Resilience

Overall qualitative definitions:

• Resilience is the ability of the system to sustain or restore its basic functionality following a risk source or an event (even unknown).
• Resilience is the sustainment of the system's operations and associated uncertainties, following a risk source or an event (even unknown)

Robustness

• The antonym of vulnerability

Safe, safety

Safe: Without unacceptable risk

Safety:

• Interpreted in the same way as safe (for example when saying that safety is achieved)
• The antonym of risk (the safety level is linked to the risk level; a high safety means a low risk and vice versa)

Sometimes limited to risk related to non-intentional events (including accidents and continuous exposures)

Security, secure

Secure: Without unacceptable risk when restricting the concept of risk to intentional acts by intelligent actors

Security:

• Interpreted in the same way as secure (for example when saying that security is achieved)
• The antonym of risk when restricting the concept of risk to intentional acts by intelligent actors (the security level is linked to the risk level; a high security level means a low risk and vice versa)

Threat

Risk source, commonly used in relation to security applications (but also in relation to other applications, for example the threat of an earthquake)

Threat in relation to an attack: A stated or inferred intention to initiate an attack with the intention to inflict harm, fear, pain or misery

Vulnerability

Overall qualitative definitions:

- The degree to which a system is affected by a risk source or agent
- The degree to which a system is able to withstand specific loads
- Vulnerability is risk conditional on the occurrence of a risk source/agent. If, for example, risk is interpreted in line with Section 1.1e), vulnerability is uncertainty about and severity of the consequences, given the occurrence of a risk source

Precautionary principle

- An ethical principle expressing that if the consequences of an activity could be serious and subject to scientific uncertainties, then precautionary measures should be taken, or the activity should not be carried out.
- A principle expressing that regularity actions may be taken in situations where potentially hazardous agents might induce harm to humans or the environment, even if conclusive evidence about the potential harmful effects is not (yet) available.

Risk analysis

Systematic process to comprehend the nature of risk and to express the risk, with the available knowledge

Risk analysis is often also understood in a broader way, in particular in the Society for Risk Analysis (SRA) community: risk analysis is defined to include risk assessment, risk characterization, risk communication, risk management, and policy relating to risk, in the context of risks of concern to individuals, to public and private sector organizations, and to society at a local, regional, national, or global level.

Risk appetite

Amount and type of risk an organization is willing to take on risky activities in pursuit of values or interests

Risk assessment

Systematic process to comprehend the nature of risk, express and evaluate risk, with the available knowledge

Risk aversion

Disliking or avoiding risk.

Technical definition: Risk aversion means that the decision maker's certainty equivalent is less than the expected value, where the certainty equivalent is the amount of payoff (e.g., money or utility) that the decision maker has to receive to be indifferent between the payoff and the actual "gamble."

Risk awareness

- Having an understanding of the risk (the risk sources, the hazards, the potential consequences, etc.)
- Being vigilant/watchful in relation to the risk and its potential consequences

Risk characterization, risk description

A qualitative and/or quantitative picture of the risk; i.e., a structured statement of risk usually containing the elements: risk sources, causes, events, consequences, uncertainty representations/measurements (for example probability distributions for different categories of consequences – casualties, environmental damage, economic loss, etc.) and the knowledge that the judgments are based on.

Risk communication

Exchange or sharing of risk-related data, information and knowledge between and among different target groups (such as regulators, stakeholders, consumers, media, general public)

Risk Evaluation

Process of comparing the result of risk analysis against risk (and often benefit) criteria to determine the significance and acceptability of the risk

Risk governance

The application of governance principles to the identification, assessment, management and communication of risk. Governance refers to the actions, processes, traditions and institutions by which authority is exercised and decisions are taken and implemented.

Risk governance includes the totality of actors, rules, conventions, processes, and mechanisms concerned with how relevant risk information is collected, analyzed and communicated and management decisions are taken.

Risk management

Activities to handle risk such as prevention, mitigation, adaptation or sharing
It often includes trade-offs between costs and benefits of risk reduction and choice of a level of tolerable risk.

Risk perception

A person's subjective judgment or appraisal of risk

Safety analysis

- Systematic process to comprehend the nature of the safety of a system and to express the safety level
- Systematic process to determine the degree of risk reduction that is sufficient to obtain a "safe system"

Public participation

A principle or practice expressing that the public has a right to be involved in the decision-making process

Risk acceptance

An attitude expressing that the risk is judged acceptable by a particular individual or group

Risk avoidances

Process of actions to avoid risk, for example, not be involved in, or withdraw from an activity in order not to be exposed to any risk source

Risk insurance

Type of insurance that is taken out against risk

Risk mitigation

Process of actions to reduce risk

Risk policy

A plan for action of how to manage risk

Risk prevention

Process of actions to avoid a risk source or to intercept the risk source pathway to the realization of damage with the effect that none of the targets are affected by the risk source

Risk reduction

Same as risk mitigation: Process of actions to reduce risk

Risk regulation

Governmental interventions aimed at the protection and management of values subject to risk

Risk sharing or pooling

Form of risk treatment involving the agreed distribution of risk with other parties

Risk tolerance

An attitude expressing that the risk is judged tolerable

Risk trade-offs (risk-risk trade-offs)

The phenomenon that intervention aimed at reducing one risk can increase other risks or shift risk to another population or target

Risk transfer

Sharing with another party the benefit of gain, or burden of loss, from the risk Passing a risk to another party

Risk treatment

Process of actions to modify risk

Stakeholder involvement (in risk governance)

The process by which organizations or groups of people who may be affected by a risk-related decision can influence the decisions or its implementation.

About the authors

Shital Thekdi is Associate Professor of Analytics and Operations at the University of Richmond. She has coauthored several papers on risk management and decision-making. She has a PhD in Systems & Information Engineering from the University of Virginia; she has an MSE and a BSE in Industrial & Operations Engineering from the University of Michigan. She has several years of experience working in the industry, with extensive supply chain management and operational analytics experience.

Terje Aven is Professor of Risk Analysis and Risk Management at the University of Stavanger, Norway, since 1992. Previously he was also Professor (adjunct) of Risk Analysis at the University of Oslo and the Norwegian University of Science and Technology. He has many years of experience as a risk analyst in the industry and as a consultant. He is the author of many books and papers in the field, covering both fundamental issues as well as practical risk analysis methods. He has led several large research programs in the risk area, with strong international participation. He has developed many master programs in the field and has lectured on many courses in risk analysis and risk management. Aven is editor-in-chief of *Journal of Risk and Reliability*, and area editor of *Risk Analysis in Policy*. He has served as president of the international Society for Risk Analysis (SRA) and chair of the European Safety and Reliability Association (ESRA) (2014–2018).

Index

Note: Page numbers in *italics* indicate a figure and page numbers in **bold** indicate a table on the corresponding page.

new risk 23, 94
new risk program 92
news agencies 34; risk communications and 53
news articles and sources 12, 14; confirmation bias and 42; credibility analysis *59*, 60–61; disinformation spread by 39; fake news 57, 60; information, analysis of reliability of 95–96; message analysis 57–58; misinformation and 57; narratives about crises and risk events 31, 101
news cycle 37
new situations: agility and 32
NHTSA *see* National Highway Traffic Safety Administration (NHTSA)
normalizing speaking up about risk 90, 91
novelty in news stories 96

oil and gas industry 10
opportunity 3, 10, 18, 29, 22, 75, 80, 86, 87, 88, 89, 93; glossary 118; loss and 78; pursuit of 70; statistical 61; *see also* new job opportunities
opportunity costs 84
outsourcing risk activities 44

Pascal, Blaise 9
peer review 45
'perfect storms' 91, 94; dependency and interdependency concepts and 22–23; financial crisis of 2008 as example of 22; key takeaway 24; risk events and 23
performance: definitions of 78; differing measurements of 79; risk and, conflict between 79
performance discipline 78
performance management and risk management, practices which support 78–84
politics 61; power politics 88
power politics 88
precautionary principle: glossary 119
preoccupation with failure 27–28
probability: classical 117; glossary 117–118; subjective 118; propensity/ frequentist 117
procedural justice 73–74
prominence in news stories 96
proximity in news stories 95
public participation: glossary 121
"putting out fires" 28, 43, 90, 92

QRA *see* Quantitative Risk Assessment
quality 79
Quality-Adjusted Life Years (QUALY) 66
quality control 10, 54
Quality Management regimes 78
quality of life 13, 66, 67
quality of work 100
QUALY *see* Quality-Adjusted Life Years
Quantitative Risk Assessment (QRA) 16

random sampling, assumption of 60
random selection 61
redundancy 31
relevance/impact of news story 96
reluctance to simplify 28
reputational damage, avoidance of 79
reputational issues 69
reputational rewards 94
resilience: commitment to 28; concept of 91, 100; dedication to 32; definitions of 12, 31; glossary 118; HRO concept of 31; plans to promote 83; risk program and 91; risk science and concept of 100; stakeholder evaluation of 13
resilience-building 43
resilient system, Hollnagel's definition of 31
risk: agreeing to disagree on uncertainty and 9–17; awareness of 74, 95, 96, 97, 99; definitions of (Appendix I) 114–115; as a good thing 75–112; three rules regarding viii–ix
risk acceptance 103; farm example **110**; glossary 121
risk amplification 22, 24, 62–63
risk analysis 43–45; concepts of 112; glossary 119; methods of 47; Society for Risk Analysis 113, 119
Risk Analysis journal, creation of 10
risk analysis plan: farm **108**
risk appetite: glossary 119
risk assessment 11; glossary 120; *see also* QRA
risk attenuation 62–63
risk aversion: glossary 120
risk avoidance: glossary 121; *see also* avoidance
risk awareness: glossary 120; *see also* awareness
risk behaviors 63
risk characterization: asking general questions to characterize risk 16; differing 11, 13–14; glossary 120; key takeaways 48; knowledge usage in

Printed in the United States
by Baker & Taylor Publisher Services